THE TRANSFORMING PRINCIPLE

The Commonwealth Fund Book Program gratefully acknowledges the assistance of Memorial Sloan-Kettering Cancer Center in the administration of the Program.

MACLYN McCARTY

THE TRANSFORMING PRINCIPLE

Discovering that Genes Are Made of DNA

A volume of
The Commonwealth Fund Book Program
under the editorship of
Lewis Thomas, M.D.

W · W · NORTON & COMPANY

NEW YORK · LONDON

Published simultaneously in Canada
by Penguin Books Canada Ltd.
2801 John Street, Markham, Ontario L3R 1B4.

Printed in the United States of America.

*The text of this book is composed in Caledonia,
with display type set in Garamond.
Composition by the Maple Vail Book Manufacturing Group.
Book design by Marjorie J. Flock.*

First published as a Norton paperback 1986

Library of Congress Cataloging in Publication Data
McCarty, Maclyn.
Transforming principle.
Bibliography: p.
1. Genetic transformation. 2. Deoxyribonucleic
acid. I. Title.
QH448.4.M33 1985 574.87'3282 84-20544

ISBN 0-393-30450-7

W. W. Norton & Company, Inc.
500 Fifth Avenue, New York, N.Y. 10110
W. W. Norton & Company Ltd.
37 Great Russell Street, London WC1B 3NU

2 3 4 5 6 7 8 9 0

To the memory of my two colleagues in the search:
Oswald Theodore Avery,
who was not inclined to write such a book,
and Colin Munro MacLeod,
who ran out of time before he could do it

CONTENTS

INTRODUCTION

ALTHOUGH EACH NEW CENTURY differs in fundamental ways from all preceding periods in history, the twentieth century seems unique for the overwhelming scale and scope of the transformation of human life affecting people everywhere. Some have turned out for the better (human health and physical ease in the affluent societies, for example), some for the worse (the fouling of our planetary environment and the massively lethal machines of warfare, for example). One way or another, most of the changes are generally believed to have been the result of science, in this most scientific of all centuries, and of the technologies that follow inevitably in the wake of scientific advance. For better or worse, our lives and those of our children's children are seen as hostages to this relatively new way of looking at and into nature, a method of thinking and working that had its beginnings only a few centuries ago and now dominates all human commerce.

The scientific method, as it is commonly termed, seems to many educated laymen an arcane, stereotyped set of rules for intellectual behavior, so specialized as to lie beyond the comprehension of ordinary people. We are in the hands of the scientists, so it goes, and we do not understand what they are up to, nor how or why they do what they do, nor most unnerving of all, can we guess what they are likely to do next.

In this atmosphere, there is need for new sources of insight into the mind of science, and into the minds of individual scientists. What motivates them and drives them along enchanted by what can in real life be the most frustrating of human occupations? How do they go about selecting the problems they wish to study? As they work, do they move from the facts at hand to hard truths, or do the facts come in as astonishments after a truth has been guessed at? Is the profession of science a self-limited endeavor, and will it, sooner or later, come to an end with all the answers in hand? Or is it, as I would guess and hope, a permanent fixture in human endeavor, likely to go on forever, each puzzle solved raising new, unpredictable puzzles.

The Commonwealth Fund, in its wisdom, has committed resources for the sponsorship of a series of books to be written by working scientists about their own work, for a general, literate readership. The books being planned (some of them already being written) will deal with the broadest range of research domains, ranging from cosmological physics and planetary biology to molecular genetics and cell biology. The writers are authorities in their various fields, caught up in the excitement of their own investigations, and capable of clear and (mostly) nontechnical prose.

This book, by Dr. Maclyn McCarty of Rockefeller University, is the first in The Commonwealth Fund Book Program series. It deals with the discovery by Avery, MacLeod, and McCarty in the 1940s that genes are made of deoxyribonucleic acid (DNA). This single discovery opened the way into the biological revolution which continues to transform our view of nature in the most intimate details, and continues as well to cast up, in its wake, one biotechnology after another for the comprehension and, it can be hoped, the reversal of human disease processes.

The selection of this book, and of those which will follow in the years immediately ahead, has been the responsibility of the Advisory Committee of The Commonwealth Fund Book Program: Alexander G. Bearn, M.D., Professor of Medicine at Cornell University College of Medicine and Senior Vice President for Medical and Scientific Affairs, Merck Sharp & Dohme International; Donald S. Fredrickson, M.D., President, Howard Hughes Medical Institute; Lynn Margulis, Ph.D., Professor of Biology, Biological Science Center, Boston University; David E. Rogers, M.D., President, The Robert Wood Johnson Foundation; Berton Roueché, Writer; Frederick Seitz, Ph.D., President Emeritus, The Rockefeller University; Otto Westphal, M.D., Director Emeritus, Max-Planck Institute for Immunobiology; and Edwin Barber, Vice President, W. W. Norton & Company, Inc. The Managing Editor is Helene Friedman.

The Editors and Advisory Committee are happy to acknowledge the constant interest and intellectual support of Margaret E. Mahoney, President of The Commonwealth Fund.

LEWIS THOMAS, M.D.

Editor and Program Director
President Emeritus, Memorial Sloan-Kettering Cancer Center
University Professor, State University of New York at Stony Brook

PREFACE

FORTY YEARS HAVE PASSED since Oswald T. Avery, Colin M. MacLeod, and I published our paper identifying the substance responsible for the transformation of pneumococcal types as deoxyribonucleic acid (DNA). Because of the nature of pneumococcal transformation, this finding carried with it the implication that DNA must be functioning as a carrier of genetic information, and the paper thus presented the first experimental evidence for the nature of genetic material. We had presented the experimental data with a minimum of interpretation and speculation, raising the question in the minds of some whether we really understood the significance of our findings. Many years went by, however, before MacLeod and I even got around to discussing the possibility of writing the story of the discovery in an attempt to clarify the sequence of events and our interpretation of them. We never got very far with this. I had too many other obligations even to contemplate the job. On one occasion, in 1969, Colin thought that he had enough free time to tackle it, and he asked me to make the laboratory notes available to him in a room at the Rockefeller Hospital. I was happy to comply with this request, which was based on his conviction that the notes should not be moved from the institution. However, he soon became involved in other activities that effectively nipped his literary efforts in the bud. When he died in February 1972, he had managed little more than a few preliminary notes. He wrote me in late 1971 that he had found it necessary to put the whole matter aside once again.

I don't believe that Avery had ever considered writing such a book. Any interest he might have had was lost long before he finished his active work in the laboratory. On the other hand, even before the DNA work had come to fruition, he liked to *talk about* the book that he could write about that "wonderful little bug," the pneumococcus, the object of most of his investigative career. The story would center on developments that came from his earlier discovery that the important capsule surrounding the pneumococcus is composed of complex sugars, and he had selected a catchy title for his imaginary book: "The Sugarcoated Microbe." I have chosen to perpetuate this felicitous designation by using it as the title of my Chapter III.

I was not able to get down to a serious effort in writing the story until after I had reached emeritus status at Rockefeller University. The question was: what kind of book should it be? There was no doubt that it had to be *my* story, since I could not speak directly for the inner feelings of my colleagues. At the same time, I felt that it was important to write it in a way intelligible to the general reader—that is, to anyone who had any interest in reading about the discovery. Soon, however, I began to have difficulties dealing in a simplified manner with the technical aspects of the research. Since I am convinced that some information about the pneumococcus (and about the biochemical and immunological approaches used in the research) is essential for an understanding of the events that led to our discovery, I have persisted in the attempt to provide what I believe to be the necessary background. I realize that there is some danger that the product of this effort will not be satisfying either to the general reader or to the biological scientist, too technical for one and too oversimplified for the other. I can only hope that the general reader, conditioned by the increased sophistication of science reporting in the daily press and in the many popular science magazines, will not be deterred by the technical detail that is included.

The young scientist is not often imbued with a sense of history relating to his activities, and indeed it has often been suggested that he should not be if he is to get on with creative work in his field. Any effort, however, to recapture the details of a path of research some forty years after it took place will soon make one acutely aware of how valuable it would be to have profuse letters, notes, and diaries of the time. I have only a modicum of these materials and have therefore relied heavily on laboratory notes and reports, which tend to be not only impersonal but usually lacking in any description of the rationale and motivation for the experiments undertaken. Any success in reconstructing the events, therefore, without resorting to guesswork or a shaky memory, has been the result of a careful review of my various sources and an analytical interpretation of the factual data and chronology. I have managed to eliminate some errors and misconceptions, but any that remain are entirely mine.

I am indebted to Miss Carolyn Kopp, Archivist, and Mrs. Sonia Mirsky, Librarian, at Rockefeller University for making available to me a complete set of the reports to the Board of Scientific Directors of the Rockefeller Institute for Medical Research from 1928 to 1948. Miss Kopp also generously provided me with a number of items from the archives relating to Avery. I am grateful to the Tennessee State Library and Archives in Nashville for giving me access to the Avery Papers in their possession.

I am happy to acknowledge the support provided by the Alfred P. Sloan Foundation in the early phase of writing this book. Throughout the later phases I have enjoyed the encouragement and generous support of the Commonwealth Fund Book Program. This has been an important factor in sustaining my efforts, and I deeply appreciate the confidence shown by the Commonwealth Committee, under Lewis Thomas, that reviewed the early drafts.

Finally, I owe a special debt to my wife, Marjorie. In addition to the encouragement and support she gave me during the writing of the book, when I was often plagued by doubts, she proved to be an extraordinarily sharp-eyed proofreader at each stage of the process.

New York 1984

THE TRANSFORMING PRINCIPLE

I

THE PREPARATORY YEARS

SURELY NO ONE is prepared to discover a faulty memory. It comes as a shock to be presented with incontrovertible evidence that one's recollection of a distant experience is faulty at best. Recently my older brother showed me a pocket diary–appointment book that my father kept in 1929, the year I started college. Among the notes that he had jotted in this book were the results of tennis and golf games with family members during the spring and summer of that year in Portland, Oregon. My father, a keen competitor, indicated the players, the winners, and often the scores of these matches, and there were numerous entries recording my participation with him in both games. The difficulty is that, while I remember the tennis clearly, I have no recollection whatever of having played golf in Portland, and this unchallengeable reminder has done nothing to rekindle the memory or to conjure up tee, green, or clubhouse.

I have had enough additional indications of these memory gaps to approach the matter of my early history and the events that determined the path of my career with some caution. I have been able to find little in the way of other diaries or documents that I can rely on to provide factual information. However, I can be certain it was sometime in childhood that I became set on the course that led me into medical science and ultimately to Avery's laboratory at the Rockefeller Institute for Medical Research. It is apparent, also, that my decision to become a doctor came too early to have been based on fully rational considerations. Although I have no personal recollection of the time that this idea became fixed, I have the assurance of my mother that it was at least by the age of 10. A few years later, when I was in junior high school, I can recall not only that I had no doubts about being headed for a career in medicine but also that in my mind this had already come to mean medical research.

The origin of my precocious interest in medical science remains obscure. I can identify nothing in my environment as a child or any specific episodes that may have given rise to it. It seems likely to me now that it came from something that I had read, but I do not have a clue as to what that might have been. Paul de Kruif's *Microbe Hunters,* a book that influenced many of my generation, was published in 1926 when I was already in high school and it thus came much too late to have initiated the process. My copy of *Microbe Hunters* was given to me by my parents for Christmas in 1929. (Actually, I had read it earlier after obtaining it from the public library.) At a minmum, the book served to reinforce my determination to prepare myself for medical research and it also led to other reading, such as Rene V. Radot's *The Life of Pasteur*.

There was little precedent in my family for a career in medicine or science, or, for that matter, for advanced education of any kind. The only exception that I know of was a maternal aunt who attended a proprietary medical school in Fort Wayne, Indiana, one of the kind that disappeared in the first quarter of this century during the revolution in medical education that followed publication of the famous Flexner report. She died tragically, before receiving her medical degree, from septicemia resulting from a cut suffered while performing an autopsy. The other members of the family on both sides were forced by circumstances to curb any educational aspirations they may have had. My mother and father were born in small villages in northern Indiana, and both had to leave school after the eighth grade to help support their families. They made the most of their limited school experience and supplemented it with reading so that their fund of information was much broader than one might have expected. My mother could recite from memory more poetry than I have ever known and she could also quote long passages from Shakespeare. She transmitted her love of literature and knowledge to her sons, and I can remember her reading regularly to my older brother and me during our preschool years. This early introduction to the world of books resulted in our becoming avid readers on our own. Our reading ranged from a series of adventure stories for boys to considerably more substantial fare, such as the books of Booth Tarkington, Mark Twain, James Fenimore Cooper, and Alexandre Dumas père. Some clue as to the timing of our selection of this reading material comes from an incident which occurred in 1920, when we were 11 and 9, relating to the naming of our newly arrived next younger brother. We were invited to suggest a name and came up with "Raoul," the hero of *The Vicomte de Bragelonne,* which we had encountered in reading this lengthy sequel to *The Three Musketeers* and *Twenty Years After*. Fortunately, it became obvious that we had no idea how the name was pronounced, and "Raoul" was dropped as a possibility.

It seems apparent now that this home environment clearly fostered the kind of aspirations that I later developed, even though I cannot detect the origin of the stimuli that generated the specific interest in medicine. My primary education was more than adequate, but perhaps a little unusual. I attended a total of five different schools in three cities before finishing sixth grade, providing a diverse experience which I believe was much more positive than negative in its impact. This view is in contrast to the one widely held by the time my own children attended school, when we frequently heard P.T.A. members express the conviction that moving from school to school was disastrously disruptive to the educational and psychological development of the child. My mother's reaction on learning of this view was to be glad that she did not know at the time that our peripatetic education was supposed to be bad for us. In retrospect, I don't believe that it was difficult to make the necessary adjustments, and we had the advantage of learning how to cope with change at an early age, as well as that of the broadening effect of varied educational approaches.

This itinerant schooling had to do with my father's occupation. He had gone to work for the Studebaker Corporation in South Bend in the early 1900s when its major activity was the manufacture of horse-drawn vehicles. The company maintained a network of factory branches for the national distribution of its products and, after he had worked his way up in the sales division, my father was sent to Portland, Oregon, when I was three years old, as the assistant branch manager. Although he was greatly attracted to Portland and the Pacific Northwest, he was also increasingly aware that the future of Studebaker lay with the automobile and he requested transfer from the horse-drawn vehicle division. As a result, in 1915 we moved to Dallas, Texas, for his first assignment in automobile sales, but in less than two years we were back in Portland where he had been appointed manager of the automobile branch.

It was in Portland on this second sojourn that I started school in 1917. When my mother took me to be enrolled in the neighborhood school it was her intention to place me in kindergarten, since she had been conditioned by her experience with my older brother in Dallas to expect regular school to begin at the age of seven rather than six. However, the school had no kindergarten, and it took some persuasion on the part of the teacher, as well as some pleading on my part, before she allowed me to remain as a first-grader. The lack of a kindergarten was symptomatic of the primitive stage of development of this new neighborhood school, composed as it was of temporary structures which for some obscure reason were referred to as "portables." My first schoolroom had six rows of desks, one for each of the half-grades from one through three, and one teacher managed the whole operation in a manner reminiscent of that described in rural schools and pioneer communities. It was almost impossible for an alert pupil not to be aware of what was going on in the higher grades, and it was an easy matter to skip a half-grade, as I did, simply by sliding over to a seat in the next row. This move had relatively little impact except to put me out of phase from then on as a mid-termer. Nevertheless, despite the questionable quality of this kind of compressed schooling by modern standards, it apparently provided me with a more than adequate grounding in the traditional three R's.

Moves to new homes served by larger, better-established schools brought me to third grade and a more drastic move back to South Bend, Indiana, and still another school where I remained through half of the sixth grade. At this point, in the summer of 1922, my father left Studebaker and accepted a new position with Nash Motors in Kenosha, Wisconsin, where I completed sixth grade in my fifth grammar school. The next stage of junior high school in Kenosha continued to provide a more than adequate academic environment along with some rather unusual special features, such as an imaginative program in music appreciation and access to a fully equipped printing shop. The latter was under the supervision of a full-time teacher who, in addition to teaching the fundamental skills of printing, directed interested students in producing the magazine for the three city junior high schools. Learning to set type, operate the press, and understand the elements of typography are among my fondest memories of this period.

Toward the end of this junior high school experience, my public school education was interrupted by a year at the Culver Military Academy, in 1925–1926. Situated in an area close to that in which my parents were raised, Culver was well known in the northern Indiana community and enjoyed a reputation for high academic standards as well as for character-building. Influenced by this, my father obviously felt that it would greatly benefit my brother and me to continue our educations in this setting. The experiment was less than an unqualified success. While I formed a poor impression of the school's vaunted academic excellence, I must confess that this view was influenced by a distaste for the traditional military academy aspects of the place, even though these were cushioned for my brother and me. We were assigned to the band, which was a separate unit at Culver, occupying its own barracks and replacing with band practice the long afternoon hours of military drill of the regular companies. My appreciation of this situation was tempered by the fact that I was not particularly adept at my chosen instrument, the trumpet. There were sixteen trumpets in the band that year. I was relegated to the sixteenth chair playing fourth-trumpet parts, which for the most part consisted of counting innumerable measures of rest and coming in—almost on time—with a single note.

There were other negative aspects to Culver, such as hazing of the first-year "plebes," a practice borrowed from West Point; but what bothered me most was the overall flavor of the place, perhaps best illustrated by the way we were greeted by fellow cadets upon arrival: "What did you do to get sent here?" This attitude and the qualities it implied weakened any appreciation I might have had for the school's merit's, academic or otherwise. In the end, a series of events, including a period when my brother went A.W.O.L., made it evident to our parents that we lacked enthusiasm for military life. We were allowed to return to the familiar setting of the public school the following year.

During my two and a half years in Kenosha High School I began to plan more seriously for training in medicine. At this time, in the late twenties, it was the perception of the lay public that the medical school at The Johns Hopkins University was *the* place that medical research and researchers were produced. I adopted the premise. This view had certainly once been valid, since in the late nineteenth century the medical school at Johns Hopkins was unique and the first in the United States with a strong scientific and academic base. However, by 1925 Hopkins had several rivals in the field. I was of course unaware of this and never considered applying to any other medical school.

Accordingly, my first move as a high school student intent on planning ahead was to send for the Johns Hopkins catalogue. From this I learned that among the requirements for admission to its medical school were a bachelor's degree and a reading knowledge of both French and German. Kenosha High School, while providing an admirable educational background in most respects, had certain limitations. It had, for example, failed to reinstitute the teaching of German after dropping it in a fit of patriotic fervor during World War I. I was sufficiently anxious about this impediment to my progress in fulfilling the Hopkins requirements that I arranged to take private lessons in German for a year from the assistant principal of the parochial German Lutheran school in Kenosha. Clearly this was unnecessary, since I would have had no difficulty satisfying this requirement during my college years, but I have had no occasion to regret the time spent in getting this firm basis in German from a knowledgeable tutor. In a somewhat similar vein, I was concerned about the mathematics curriculum at the high school, which did not go beyond advanced algebra, so I was motivated to arrange for trigonometry lessons during the summer from one of the teachers at the school.

In retrospect this behavior strikes me as overzealous and suggests that I may have had some of the attributes of an intolerable grind. However, this is an inaccurate picture, since I had many other interests and found time for frequent tennis and swimming, for playing in the high school band (with, alas, no notable increase in skill over my efforts at Culver), and for dating, dances, and parties. In fact, from my present vantage point over fifty years later, I find it difficult to believe that my friends and I were able to pack all of this activity into our daily lives, a reaction that no doubt merely reflects the usual lack of comprehension on the part of the aging for the energy of youth. I had also set up a chemistry laboratory in my basement, and three of my classmates, each with his own home lab, joined me in forming a club that we called the Amateur Research Chemists. The key word here is "amateur," since we were certainly inexperienced and generally incompetent as chemists, creating some hazard to ourselves and others. As a result of our access as alumni to that junior high school printing press, we even had a letterhead, with a heading in Gothic type proclaiming "The A.R.C. Club" and a definition of the initials in parentheses below, but no address other than "Kenosha, Wisconsin." We all had chemicals which we probably shouldn't have had—I recall my bottles of both sodium and potassium metal stored under kerosene—and one of my friends managed by the use of this letterhead to acquire things from J. T. Baker Chemical Co. that I doubt he could have obtained without it. One of my memories of this period is of spilling liquid bromine on his head while I was in the process of trying to open his newly acquired bottle of this corrosive element. This was quickly washed away without harm, and in general we survived our inept excursions into laboratory science, although I did sustain a rather severe burn of one hand from hot sulfuric acid. These experiences contributed to my learning—the hard way—that maintaining laboratory safety is one of the first principles of research.

In contrast to the clear recollection of the reasons for the early selection of Johns Hopkins for my medical training, I am totally blank on the subject of what led me to choose Stanford University for premedical education. It certainly had nothing to do with the family move back to Portland, Oregon, for the third time in March 1929, one month after my graduation from high school. My application to Stanford—I overconfidently filed no other—had gone in before I knew of the decision to move. That decision on the part of my father to return to his beloved Portland proved to be badly timed, and by the end of my first year at Stanford we were on our way back to Kenosha again. Portland had become economically depressed even before the October 1929 stock market crash and, finding little in the way of opportunity for his contemplated new business activity, my father had become increasingly receptive to Nash's offer to return as vice-president and general manager, with the clear understanding that he would become president in the near future.

Whatever considerations led me to Stanford, the result was fully satisfactory. The school had not attained its current national reputation and was populated principally by Californians. But it had high academic standards and a distinctive flavor which set it apart from schools on the East Coast. The unique architecture of its buildings, which covered only a small segment of its extensive campus, together with the benign climate and unusual vegetation, contributed to this flavor. There was also, however, an informal air and a sense of personal freedom that reflected the university's German motto: *Die Luft der Freiheit weht.* All in all, it seemed a felicitous environment for the next stage of my preparation for medical science.

Within the limits imposed by the Stanford curriculum for students in the "Lower Division"—that is, freshmen and sophomores—I still kept my eye on those requirements for admission to The Johns Hopkins Medical School, getting an early start on more French and German as well as on mathematics and the biological and physical sciences. From the point of view of preparing for future research on the pneumococcal transforming substance, the most important step was a second-year decision to major in biochemistry. Biochemistry at Stanford in those days bore little resemblance to the large and diverse departments in universities today. It was a part of the Chemistry Department, not a separate unit, and staffed by a single faculty member, Professor James Murray Luck. All of the lectures in the general biochemistry course, which was required for the first-year medical students who consituted most of the class, were given by Luck, and he ran the associated laboratory course with the aid of one graduate assistant. My additional activities as a biochemistry major included special experiments in the laboratory, an evening seminar held every week or two at Luck's home in Palo Alto, and finally a full-time research project during the last quarter of my senior year.

Dr. Luck at this period was launching the *Annual Review of Biochemistry*. Its first volume appeared in 1932. This series remains an important reference source in the field today, and furthermore it spawned the development at Stanford of annual reviews in more than twenty other fields of science. Thus, it has become a major enterprise in scientific publishing. Early on, Luck would carry galley proofs of the reviews in his pocket, and more than once he gave me marked sections of the galley to indicate the references that were to be my assignment for reporting at the seminars. I have kept my copy of Volume I of the Annual Reviews for its historical interest. In the context of the DNA study, the most interesting feature of this seven-hundred-page volume is the almost total absence of any discussion of the nucleic acids. The study of this component of living tissues simply was not a popular field of biochemistry fifty years ago.

My first real taste of laboratory research came toward the end of my last year at Stanford. The project that Luck outlined for me had to do with liver proteins. Specifically, did the considerable increase in the size of the liver in animals fed on high-protein diets involve storage of specific proteins or simply overall growth of the organ? The experimental animals were rats maintained for a few weeks on either high- or low-protein diets and then sacrificed so that the liver could be removed, extracted, and separated into fractions for analysis of protein components. I can't say that I was greatly intrigued by the problem even at the time, but it did provide a considerable breadth of experience in laboratory activity. I prepared the special diets, tended the rats during the experimental period, carried out the operative procedures that included perfusion of the liver with salt solution to remove excess blood, and then proceeded to the extraction, fractionation, and analysis of the material. The poor definition of the separated liver protein fractions was probably the weakest aspect of the study, depending as it did on differential solubility properties. The most demanding part of the research was analysis of the protein content of each fraction by the Kjeldahl method. This involved digesting the samples in strong acid to convert the nitrogen of the protein to ammonia, followed by distillation of the ammonia and its accurate quantitative measurement by a titration procedure. My difficulties in setting up the array of apparatus for these analyses left me with a strong distaste for the Kjeldahl method which I have never lost.

This was my first experience with laboratory animals, and the rats and I did not become completely comfortable with one another in the course of the project. The cages were quite deep, making it necessary to reach to the back rather blindly at arms length in order to grab an animal with a gloved hand. Naturally resenting this intrusion, the rats tended to counter by biting a gloved finger. The gloves prevented any real damage, but I generally emerged from these encounters so shaky that it was difficult to get on with the operative procedures, particularly the perfusion of the liver, which called for a steady hand to get a needle into the small portal vein.

The outcome of this research indicated that all of the several poorly defined protein fractions were increased in livers of rats on the high-protein diet and thus suggested that growth of the liver rather than simply storage of protein was involved. However, it was obvious that in only ten weeks of that final quarter at Stanford I had not completed the type of well-controlled experiments required for definitive answers. Dr. Luck had departed before the end of the quarter for a summer at Cambridge University, where he had received his early training. I was to mail him a report of the work when I was finished. I still have my copy of this report, and a thin and amateurish document it is; but at least it does not go beyond the limited data in reaching conclusions. Luck later returned to this project in a more extensive and sophisticated study that he published in *The Journal of Biological Chemistry*[1] in 1936. When I saw this paper I derived some comfort from the trend of the overall results. They did not differ greatly from those of my preliminary and incomplete efforts. The experience, despite its frustrations, had at least done nothing to shake my determination to pursue the goal of medical research.

Having not lost sight of this goal, I completed my application for admission to the medical school at Johns Hopkins during the Christmas holidays in 1932 and received acknowledgment of its receipt in early January. Dr. Luck tried to persuade me to alter my plans and apply to Stanford Medical School, but I had encountered nothing that weakened my determination with respect to Hopkins and ended by submitting again only the single application. Late in January 1933 I was asked in a letter from the assistant dean at Johns Hopkins, Dr. E. Cowles Andrus, to "arrange for an interview with the regional representative of the Committee on Admissions, Dr. Emile F. Holman, Stanford University Hospital, San Francisco, California." Dr. Holman was professor of surgery and a Johns Hopkins graduate who had trained under the great Halsted, one of the "big four" of the early years of the Hopkins medical school. My appointment with Dr. Holman was at 3:30 in the afternoon on February 8. Despite the hour, it appeared to fall in the middle of his operating schedule since he appeared in his office for the brief interview in a surgical scrub suit. He was friendly enough to allay my mounting nervousness, intensified by the sense of interrupting a busy schedule, but I left with no idea whether I had made a favorable impression. Fortunately for my peace of mind, things moved rapidly in the Hopkins admitting office in those days. A letter dated February 21, 1933, from Dr. Andrus told of my admission "to this Medical School contingent upon your completing the courses which you are now pursuing at Stanford University." Having cleared this important hurdle, I faced the final months at Stanford in a relaxed mood.

II

THE MEDICAL SCENE

After a largely midwestern upbringing and four years at Stanford, arrival at the East Baltimore setting of The Johns Hopkins University School of Medicine gave rise to something like culture shock. I had never seen anything like it. In every direction from the school the streets presented a line of attached row houses. All boasted the same simple design in red brick, built flush with the sidewalk and unrelieved by the softening influence of lawns and trees. I knew no one in Baltimore or in my incoming class at the medical school with whom I could seek diversion during those first few days to shake off the depressing effect of the strange environment. There were no dormitories for medical students, and my rental room on the third floor of a row house on North Broadway did not ease the depression. An episode connected with the renting of this room pointed up some of the more parochial aspects of my early experience despite the moving about that I had done in the West and Midwest. After I decided to take the room on a late September afternoon, I tried to pay my new landlady the agreed-upon monthly amount in advance. She declined the cash, saying that she could not accept money on that day because it was Yom Kippur. I was totally mystified, having never heard of this or the other Jewish holy days and knowing nothing of the rules for their observance. I was clearly about to have my horizons broadened.

Eating was also a problem in East Baltimore, which offered little attraction to the enterprising restauranteur. Although I had arrived determined not to get involved with a Greek-letter society as I had at Stanford, the blandishments of the medical fraternities, for nothing else than to serve as an eating club, became hard to resist. My doubts were sufficiently deep-seated, however, that I consulted the dean of the school for advice and reassurance before joining a fraternity. The activities of the fraternities in their quest for new members had the salutary side effect of getting the new students acquainted with one another, and in that manner I found a group of congenial colleagues who helped dispel my black view of East Baltimore.

It was at one of the "rushing" affairs of a medical fraternity that I got my first indication that Johns Hopkins was not quite the ivory tower of medical research that I had fondly supposed it to be. A graduate member of one of the fraternities, then a resident in gynecology, who had written to me during the preceding summer on behalf of his fraternity, engaged me in conversation about my plans for the future. When I told him that I planned to go into medical research, he laughed and responded: "That's what they all say. You'll change your mind before you finish." Although I considered this a frivolous and unreliable comment, there is no doubt that there was more than a germ of truth in his implication that the emphasis on research that I had anticipated no longer existed. Only two or three members of my class of seventy have devoted their careers to research.

The most effective antidote for my distaste for the new environment in which I found myself was being immersed in the demanding curriculum of a first-year medical student. Gross anatomy, the dominant part of the beginner's course, soon presented a real challenge because of the inherent difficulty of sorting out the complex interrelationships between the numerous structural components of the body, all with unfamiliar Latin names. One could not fully understand the course and distribution of a nerve or blood vessel without at the same time knowing something of the organs it served and the structures through which it passed. This dilemma sent the student ranging through the anatomy textbook from one section to another in a frantic attempt to get some kind of integrated picture. It is not easy to find much in the way of intellectual stimulation in the study of anatomy, but there is no doubt that it supplies information essential for dealing with the more dynamic problems of function.

My principal relief during this first term from activities such as dissecting a cadaver and poring over anatomy textbooks and atlases came from the opportunity to resume the study of biochemistry that I had begun at Stanford. At Hopkins the department and the course were called "Physiological Chemistry," but this represents only the most trivial of the several differences between the teaching of biochemistry at the two schools. There were at least three full-time and relatively senior faculty members in addition to the head of the department, Dr. W. Mansfield Clark, as well as a number of graduate students; and teaching was definitely not a one-man show as at Stanford. Dr. Clark clearly put his stamp on the character of the course, however. His emphasis on the physical-chemical aspects of the subject resulted in the presentation of a picture of biochemistry which was completely different from the one I had gotten at Stanford. Nevertheless, the two courses complemented each other in a way that increased the breadth of my exposure to the subject.

In recognition of my previous experience at Stanford, I was not required to participate in the laboratory exercises in physiological chemistry but was permitted instead to undertake a special laboratory project under the direction of one of the professors, Dr. Leslie Hellerman. This could hardly be called research, since the assignment that Dr. Hellerman gave me was to try to repeat some experiments on the purification of heparin, the blood anticoagulant, that had recently been published in *The Journal of Biological Chemistry.*[1] Nonetheless, the techniques and procedures required in the work were sufficiently diverse to expand my laboratory experience. The experiment began with the extraction of crude heparin from beef liver—liver seems to have been the dominant theme of my early laboratory experience, but I have scarcely touched it since, even as food—and the original workers dealt with one-hundred-pound lots. Since we did not have the facilities for working on this scale, I limited my efforts to tackling twenty pounds of the organ at a time, which was quite enough. Some weeks and numerous purification steps later, I wound up with a small amount of white powder which still smelled remarkably like liver but had a fair amount of anticoagulant activity. It did not, however, have the crystalline properties that the original authors had described for their preparations. In later years I labored under the inaccurate impression that they had claimed in their papers to have crystallized heparin, and I used this on occasion as an example of how skeptical one has to be of published results, even when they appear in the best scientific journals. Heparin has not yet been crystallized. Nor is it likely to be, because it is now known to belong to a class of substances with the technical name of sulfated glycosaminoglycans, which are not amenable to crystallization. On reexamining these papers almost fifty years later, I find that the authors had carefully avoided this claim and cited reasons to doubt that their crystals represented heparin itself. The lesson I learned about scientific skepticism was thus supplanted by another: it is important to check one's facts before drawing conclusions.

This heparin episode was my last fling at extracurricular laboratory activity for some time to come. I did not seek opportunities to engage in research projects during freetime and summer vacations. Such opportunities were quite rare anyway in the pre–World War II medical school but also I had made the decision to restrict my efforts to completing the regular course work and to come back to research on a full-time basis once this was done. The demands of the medical curriculum were great enough to justify this as a realistic approach. Hopkins had a long tradition of having no examinations or quizzes during the first two years (and of supplying no grades to the students at any time). During one week at the end of the second year comprehensive written *and* oral examinations were held in all of the preclinical sciences: anatomy, biochemistry, physiology, pharmacology, bacteriology, and pathology. This struck me as an advanced and sensible way of conducting the business of education, but no doubt it imposed an intolerable stress on certain members of each class; largely for this reason, I believe, the system was abandoned the year after our class completed the exercise. Our clinical courses during the next two years were also given without formal examinations until the end of the last year, but this seemed to create fewer problems for us, probably because the subject matter was more pragmatic and less theoretical. A record of our performance in each course, in lieu of grades, was kept on file in the dean's office, closely guarded as confidential data. This posed some difficulties for the active members of the honorary medical society, Alpha Omega Alpha, when they met each year to elect new members from the senior class. A senior member of the faculty had to attend with copies of the necessary records. As each name was proposed he would indicate without divulging any details whether the candidate's academic record justified his being considered for membership. My election to Alpha Omega Alpha was about the only measure I had of my performance.

Since I had started Stanford less than a month before the Black Tuesday debacle in 1929, essentially all of my premedical and medical education came during the Great Depression. I was sheltered from its impact and never threatened with interruption of my schooling through lack of financial support from my parents. Furthermore, I rarely read the newspapers deeply or discussed economic or political topics with my colleagues. As a result I was only vaguely aware of the overwhelming impact of the economic collapse, and later almost equally detached from reports of the growing clouds of war in Europe. It was not that I was unconcerned, simply poorly informed and preoccupied with other things.

The Hopkins students who did not have to work during the summer holidays in the mid-thirties, or those who were not lucky enough to find jobs, had the choice of relaxing to recover from the rigors of the previous year or of seeking some activity to supplement their medical education. I did a little of both. I began the summer after the first year by getting married. The event underscores the generosity of my parental support and placed me in an unusual position with my fellow students. Married medical students are commonplace today, but there was only one other married member of my class at the time. Aside from getting me out of East Baltimore, since I had no intention of asking my wife to live in that area of the city, I do not believe that my marriage changed the course of my medical training. Being unaware that the authorities took a dim view of early marriages, particularly at the postgraduate level of internships and residencies, I had asked no one's permission. Still later I found that there was even more explicit disapproval at the Hospital of the Rockefeller Institute for Medical Research. Both the original director and his successor, who held the post until 1955, subscribed to the view that the aspiring medical scientist should defer marriage until he was well established, being wedded in the meantime to the research laboratory. However, this general attitude was beginning to erode even before the war, and my marital status never proved to be a handicap for me.

After a honeymoon trip by automobile that summer, I also drove to Rochester, Minnesota, intending to find some medical activity for vacation time the following year. I found that the head of surgical pathology at the Mayo Clinic, Dr. William Carpenter MacCarty (no relation, but a Johns Hopkins graduate), had the custom of taking on a group of medical students each summer for experience in his specialty. He agreed to include me in the next group. Again I had done this without consulting anyone at Hopkins, and I was shaken a little by the reaction of my examiner in pathology during the hectic week at the end of the second year. My oral examination was given by the head of the pathology department, Dr. William G. MacCallum, a senior and highly respected member of the faculty, who had been at the school since its early, heroic days. In the course of quizzing me, he inquired about my plans for the summer; after I had told him, he remained silent for a few moments, contemplating the ceiling, and then said, "Well . . . you will probably see a lot of material and gain some experience. Just don't believe everything you hear." I discovered later that the reason for this caveat was MacCarty's unorthodox views about certain aspects of pathology. Most notably he had claimed that it is possible to diagnose cancer on the basis of the appearance of single cells in sections of frozen tissue. MacCallum was right, however, about being exposed to a large amount and a variety of material. Armed with his warning I had an instructive summer at the Mayo Clinic without becoming a disciple of MacCarty's views. MacCarty was a warm and generous man, and he gained the respect of all the members of the group of medical students that he voluntarily provided with an opportunity for additional experience.

As I got into clinical studies the following year, I began to believe that emphasis on pediatrics would be the best way to promote my interest in the infectious diseases. This interest had its roots in my earlier reading, such as *The Life of Pasteur* and *Microbe Hunters*, but it had certainly not been enhanced by the second-year course in bacteriology. Primarily a dry, systematic presentation of the subject, dwelling on the classification and methods of identification of the principal classes of pathogenic bacteria, the course did little to highlight the glamorous aspects of the subject. Having said this, I must also concede that I first heard about the transformation of pneumococcal types from one of the lectures of the professor of bacteriology. His description of the phenomenon was so graphic that I retained it until I came face to face with the reality in the laboratory several years later.

In any event, I followed through on my notion by spending part of the summer after the third year as a substitute intern in pediatrics. This kind of substitution is common and satisfies both the needs of the medical student for some practical experience and the requirements for filling house staff positions during vacations. In my case, it provided just the exposure I needed to settle on pediatrics as the area to round out my medical training. The most prestigious internship at Johns Hopkins was on the Osler medical service, and the general esteem in which it was held led me to apply for this as well as for pediatrics. However, as the fourth year progressed, I realized that I was being influenced by irrelevant considerations and withdrew my application for an intership on the medical service, leaving myself as usual with all of my eggs in one basket. It was a boost to my morale to have the professor of medicine, Dr. Warfield T. Longcope, stop me in the hospital corridor shortly thereafter to ask me why I had withdrawn. I would not have guessed that he even knew who I was.

It is legitimate to inquire why I bothered to take time for hospital training in the first place, since I was planning a career in laboratory research. This is an old question which has been answered differently by different people. As a matter of fact many successful medical researchers have opted to go directly into the laboratory after receiving their medical degrees. In my own case, I felt that the considerable body of theoretical and systematic knowledge obtained during the medical school years required supplementation with practical experience in order to make it complete. In addition, I must confess that I looked upon the clinical training as a hedge. I might not make a go of it as an independent investigator and I wished to be ready for medical teaching as a fall-back position. Medical practice was a poor third among the possible options. In any event, I have never regretted the three years that I spent as a pediatric house officer. I am still convinced that this kind of postdoctoral clinical experience is an important part of the training in human biology for anyone who expects to engage in disease-oriented research.

The late 1930s were exciting years for a young student of infectious diseases. The sulfonamide drugs, just introduced, had ushered in an era of rapid change in our ability to deal with bacterial infections. I had enough experience as a medical student with the pre-sulfonamide problems of treating severe infections to fully appreciate the power of first the sulfonamides and later penicillin and a long series of antibiotics. One memory lingers from that summer of 1936. As a substitute intern in pediatrics I participated in futile efforts to save a child with streptococcal meningitis, a disease that occurred occasionally as a complication of middle ear and mastoid infection. It was uniformly fatal. Early in the period of my regular internship the following year we were able to treat a similar case with sulfanilamide. It is easy to understand the sense of elation we all felt when the boy promptly began to improve and ultimately recovered completely. In our concern today about the misuse of antibiotics and the emergence of antibiotic-resistant strains of bacteria, we tend to forget that there are a number of infections, like streptococcal meningitis, from which no one had recovered prior to the discovery of effective antibacterial agents.

There was probably no better place for clinical training in pediatrics than the Harriet Lane Home for Invalid Children, as the pediatric unit of The Johns Hopkins Hospital was then known. All of the essential elements were there—a wide selection of case material that posed the full range of disease problems; a superb full-time faculty ably supported by a conscientious group of part-time practitioners; a house staff that was drawn from the top medical graduates at various schools across the country; a nursing staff that worked in unison with the physicians in the care of patients; and a social service unit that was remarkable in its day for its breadth of coverage and effectiveness. The moving spirit in the creation of the aura of excellence and harmonious cooperation that pervaded the Harriet Lane was the chief of pediatrics, Professor Edwards A. Park. Dr. Park was a tall, Lincolnesque figure, soft-spoken and gentle, whose considerable erudition was often obscured by his self-effacement. His humility was real, and he seemed unaware of his ability to bring out the best in his associates at all levels or of his own major role in creating an illustrious clinical setting. He was research-oriented himself, devoting much of his active career to a study of rickets and scurvy. Thus, he was most sympathetic and encouraging to those who aspired toward research. He gave me a helping hand on repeated occasions and continued his interest in my progress well into his retirement years.

For those interested in infectious diseases, Harriet Lane had another resource in the bacteriology laboratory under the able supervision of Miss Helen Zepp. Zeppie, as we called her, had little patience with ineptitude. In fact, she terrorized some of the interns by her vocal criticism of their bumbling efforts in handling bacterial cultures, but she was an excellent teacher and guided those of us who persevered in the intracacies of practical bacteriology. It was under her tutelage that I got a more realistic grasp of the subject, and I spent many of my "free" hours in that laboratory.

The bacteriological laboratory was the scene of my contribution to a study of pneumococcal pneumonia that led to my first published paper. This took place during my second year at Harriet Lane when I was an assistant resident. The first modified sulfa drug, sulfapyridine, had just been introduced with the property—which sulfanilamide lacked—of being effective against the pneumococcus. It seemed worthwhile to carry out a test of its effectiveness in the treatment of pneumonia in infants and children. The study was under the direction of Dr. Horace Hodes who had recently returned to Baltimore after two years of training at the Rockefeller Institute to serve as director of the city's infectious disease hospital and as a staff member at Harriet Lane. In addition to participating in the clinical management of the pneumonia cases, I did part of the bacteriological work involving isolation and serological typing of the offending pneumococcus in each case. The effectiveness of sulfapyridine was abundantly evident in the seventy-one cases included in the study. My colleagues and I reported the results in a paper published in the *Journal of Pediatrics*[2] for which I prepared the illustrative charts as the amateur medical illustrator of the group. This clinical study was an important episode in my training, since it gave me the opportunity to become familiar with the properties and methods of handling of the microorganisms that would later play a central role in my research.

My development as a pediatrician was certainly facilitated by having our own young children to observe and learn from. Our older son had been born just before I started the third year of medical school, and his younger brother came along toward the end of my intern year. Aside from the pleasure and enrichment of our lives that the boys provided, they also gave me a much better feeling for some of the problems presented by parents who brought their children to the Harriet Lane clinic. Not that I was given much time to indulge in simple enjoyment of my family. The clinical responsibilities of the house staff were demanding, requiring long hours with little scheduled time off, but the rewards were great. Dr. Park gave the members of his resident staff a large measure of independence, relying on their judgment and on peer interaction to move them to seek assistance from the faculty and attending physicians when it was needed. The assistant residents made decisions regarding admission to the hospital, an onerous responsibility that at the same time provided a strong stimulus for the development of clinical judgment and acumen. I found my activities as assistant resident enjoyable and absorbing so that I was happy to stay on in this capacity for a second year, even though I was still looking ahead to the adventure of laboratory research. I decided that this would be enough time to devote to clinical training and began early in the final year to look for an appropriate laboratory position.

There were limited opportunities for postdoctoral research experience for M.D.s in those years just prior to World War II. Only a handful of postdoctoral fellowships were offered, and not many laboratories had the resources to accept and support trainees on their own. An exception to this general state of affairs was found in the Rockefeller Institute for Medical Research where the individual laboratories had line items in their budgets for the support of young hopefuls seeking research training. Dr. Park seemed to have a good line of communication on the situation at Rockefeller. He had managed to place several of his research-oriented house officers there in the past. In the fall of 1939 he told me of an opening in the laboratory of Dr. Leslie T. Webster at Rockefeller and suggested that I apply for it. It was thus that I made my first visit to the Rockefeller Institute where I was greeted and shown around by a former assistant resident at Harriet Lane who was about to leave Webster's laboratory for the University of Chicago. I had already been duly impressed by the reputation of the Institute and now found myself somewhat awed by its outwardly austere atmosphere. Dr. Webster was friendly and sympathetic during our interview, but in the end it was clear that I did not fulfill his criteria for the position. He was planning to embark on a study of the effect of diet on infection and was looking for someone trained in the science of nutrition.

Disappointed but not discouraged, I returned to Harriet Lane for more help from Dr. Park. Not long afterward, he had information on some other possibilities, one of which was with Dr. William S. Tillett at New York University. Dr. Tillett had been on the Hopkins faculty when I was a medical student, and I remembered him for his informative and stimulating lectures in the area of infectious disease. He had left Hopkins in 1937 to become the professor of bacteriology at New York University. The following year he became the first full-time chairman of the department of medicine of that school. Although I was unaware of the details of his scientific work, I knew that he had an excellent reputation as an investigator. The idea of working under him appealed to me, and I made another trip to New York, this time with greater success. In a letter sent January 15, 1940, Tillett wrote me as follows:

Dear Dr. McCarty:

Concerning the position here which we have discussed, I find that I am able to offer you a position as a Fellow in Medicine for the academic year beginning July 1, 1940, at a salary of $100 a month. A desk in my laboratory, together with its facilities, will be available for you. The exact nature of the problem we can decide on at leisure.

If you finally decide that you would like to take this position, I will be delighted to have you and hope it will be an interesting and pleasant experience.

Sincerely yours,
W. S. Tillett

I promptly accepted saying that "I believe the fellowship with you offers me exactly what I wanted in the way of trying my hand at laboratory work. . . ." The "$100 a month" was not an unusual compensation for a scientific trainee in the prewar period. Since the resident staff at Johns Hopkins received nothing but room and board (which as a married house officer I didn't use), I was not dismayed by the figure. However, I doubt that even a single person could have lived comfortably on that amount in New York, even in 1940, and so I was destined to continue to be dependent on the largess of my parents.

Although I left Harriet Lane with some regrets, my departure from Baltimore itself in late June 1940 was most welcome. I had never developed any real fondness for the city, and my duties as a pediatric house officer had required that I move back to East Baltimore in the immediate vicinity of the hospital. In preparing for the move to New York we had found a house with reasonable rent in an old residential area of Flushing on Long Island. It was well enough endowed with trees and a small lawn to satisfy the expectations generated by my midwestern upbringing, and it provided a more acceptable setting for raising our two young sons. We were quickly settled in this new home and I was ready to join the ranks of the subway commuters.

My official letter of appointment from the secretary of New York University earlier that June had indicated that I was to begin on September 1. I ignored this and presented myself at Tillett's laboratory on July 1, only to find that he also was not expecting me until September, despite the date given in his letter offering me the position. He was about to leave for Maine for the summer. Even so, I could use the laboratory during that time, minus his supervision and the $100 a month. I was too eager to start to let this opportunity go by, making use of the time for preliminary experiments and a lot of necessary reading in the library. His laboratory was on the same floor as the department of bacteriology in an old building at the corner of First Avenue and Twenty-sixth Street, across from Bellevue Hospital. The whole area was somewhat seedy and there was little of that atmosphere of outward elegance that had impressed me on my visit to the Rockefeller Institute. However, the laboratory itself, located in the back of the building next to the animal quarters, was more than adequate for my needs.

As a rather natural outgrowth of my recent clinical experience, my initial laboratory efforts dealt with the sulfonamide drugs. One of the developments that had caught my eye was described in some recent reports from England. The rather simple organic chemical *p*-aminobenzoic acid was apparently capable of completely inhibiting the antibacterial effect of sulfanilamide both in the test tube and in experimental infections. This provided a clue as to the mode of action of these new drugs. Nothing much came of my efforts to exploit this finding except for a small paper showing that *p*-aminobenzoic acid also nullified the curative effect of sulfapyridine in pneumococcal infections of mice but did nothing to alleviate the toxicity of the drug.[3]

After Tillett returned in the fall, he suggested a project involving the role of white blood cells (or leukocytes). Leukocytes were of known importance in defense against bacterial infection through their ability to engulf and destroy the microorganisms, but were they essential for the curative action of the sulfonamide drugs? The protocol called for the use of rabbits which were to be rendered deficient in leukocytes by the only method then available—the administration of benzene. Once the animals had become sufficiently deficient in leukocytes—a state referred to as leukopenia—they were to be infected with pneumococci and then tested for the therapeutic efficacy of sulfapyridine in comparison with a group of control animals with normal white blood cell counts. I carried out all of the procedures, including the administration of the drugs, the total and differential leukocyte counts, and the bacteriological work.

A number of difficulties arose that made it impossible to achieve the original goal of this project. The most important of them was the finding that sulfapyridine appeared to reverse the leukopenic effect of benzene. As a matter of fact, we found that if benzene and sulfapyridine were administered concurrently the fall in leukocyte count was totally prevented. The work of toxicologists on the action of benzene, an industrial poison, had indicated that the toxic effect was brought about by some product of the oxidation of benzene in the body rather than by benzene itself. This suggested that sulfapyridine might be acting by interfering with the oxidation of benzene in the tissues. I got a chance to try my hand at some biochemical experiments again in showing that sulfapyridine did indeed markedly suppress the amount of oxidation products (phenols) excreted by animals receiving benzene. Dr. Tillett and I submitted this work for publication early the following summer.[4]

As this work was getting started in October 1940, I wrote to Dr. Park giving him an account of my progress and describing my activities in Tillett's laboratory. He replied with the first of a series of invitations to return as chief resident physician at Harriet Lane. After the invitation, he added with characteristic diffidence and honesty the following:

> I cannot in all honesty advise you to be resident. What you really ought to do in your own interest is to go on in bacteriology for at least one more year and perhaps for two years. It would be splendid if you could spend on year at the Rockefeller Institute. If I can help you in furthering your plans for the continuation of your bacteriological studies, please let me know and point out the way I ought to proceed.

This advice agreed with my own inclinations, since I had no intention of leaving laboratory work until I had given it a thoroughgoing try and then only if I failed at it. On November 9 Dr. Park renewed his offer in a letter that crossed with mine telling him of my decision to stay with Tillett. He more or less reiterated his previous comment by saying:

> I cannot urge you to be resident and believe that if you go on for another year with Tillett or better at the Rockefeller Institute, it would be the best plan for your future.

In our discussions of plans for a second year, Tillett had suggested that I apply for a National Research Council Fellowship in the Medical Sciences in the hope of obtaining a somewhat more adequate stipend. I submitted the application before Christmas in 1940, including Dr. Park among the names of those who would serve as references. I gave little further thought to this and was deeply immersed in the work on sulfapyridine and benzene that I have just described when, in March of 1941, I received a letter from Dr. Francis G. Blake, the chairman of the Medical Fellowship Board, notifying me that I had been awarded a fellowship beginning September 1, 1941, "with a grant of $2,300." The letter then added this bombshell in the next paragraph:

> The Board would like to suggest that you give consideration to the possibility of working in some other laboratory than that indicated in your application; for example, with Dr. Colin M. MacLeod at the Hospital of the Rockefeller Institute, with a view to broadening your experience. I hope that this suggestion of the Board will appeal to you as a desirable one. Will you kindly let me know as soon as possible whether you are willing to accept this suggestion and, if so, where and with whom you would like to work.

Because of Dr. Park's repeated emphasis in his letters to me on the desirability of going to the Rockefeller Institute, I have had occasion to wonder whether he had written anything in his letter of recommendation that had influenced the Board to make this suggestion. I never found the answer to this. In any event, when I showed Blake's letter to Dr. Tillett, after an understandable first reaction when he said, "I wonder if this is a crack at me?", he immediately began to take action to make the necessary arrangements. He happened to know that Colin MacLeod had accepted the position of chairman of the department of bacteriology at New York University and on July 1 would be leaving Avery's laboratory at Rockefeller and moving to the same floor of the building in which I was then working. Tillett had maintained a close personal friendship with Avery dating back to the 1920s when he had spent several years at Rockefeller in his research group. He lost no time in picking up the telephone to call Avery and asked him if he would accept me as a fellow in his laboratory. Clearly acting on the basis of Tillett's recommendation, since he hardly knew me, Avery agreed. The approval of Dr. Thomas M. Rivers, director of the Rockefeller Hospital, as well as approval of the Fellowship Board were quickly obtained. I was set on a course for Rockefeller.

I promptly wrote Dr. Park to convey the good news, and

he replied with a congratulatory letter, including some of his characteristic comments:

> I am glad that you are going to work with Dr. Avery at the Rockefeller Institute. Everyone seems to think he represents the extreme upper stratosphere. I hope that you do not develop "Bends" in making the ascent.

Dr. Park knew Avery from his early years, having been his medical school classmate at the College of Physicians and Surgeons of Columbia University. I learned from him later that he had not considered Avery one of his more impressive classmates and had always been surprised by the outstanding reputation he had acquired.

The role that chance plays in shaping one's career is clearly evident in the course of events leading me to the Avery laboratory at the right moment. The sequence that I have recounted could have been altered at several points, resulting in a different outcome. If Dr. Webster had accepted me at Rockefeller on that first attempt, if I had chosen one of Dr. Park's suggested mentors for research training other than Dr. Tillett, or if my National Research Council fellowship had not been timed to coincide with Colin MacLeod's departure from Rockefeller, it is unlikely that I would have come to study the substance responsible for the transformation of pneumococcal types.

I had met Dr. Avery at a dinner in Tillett's home and found him charming and a fascinating raconteur in this social setting, but I was still not thoroughly acquainted with his scientific contributions. When I visited him in the spring of 1941 to discuss plans for the coming year, he treated me to one of his famous monologues describing the earlier work of his laboratory on the pneumococcal polysaccharides. Nothing specific was mentioned about the project that I might undertake, although he gave me a number of reprints of his scientific papers and other reading material to peruse over the summer. He did not touch upon the work on the transformation of pneumococcal types. Since there had been no publication on the subject from the laboratory since 1934, I remained ignorant of his recent studies with MacLeod on transformation until I arrived to begin my fellowship in September.

Thus, I come to the point where my part in the story of the discovery of the genetic role of DNA begins. However, in order to provide a basis for comprehending the research and its significance, it will be necessary first to introduce the reader to the pneumococcus in a more intimate fashion and to describe the investigations that paved the way for our later work. This will be the object of the next three chapters.

III

THE SUGARCOATED MICROBE

IT IS OFTEN POINTED OUT that research in the basic sciences provides the base of new knowledge essential for the development of the applied sciences, including medicine. We are less frequently reminded that the reverse can also occur. Research directed against a specific medical problem has resulted in contributions to fundamental biological knowledge. The most dramatic example of this is the discovery that deoxyribonucleic acid (DNA) is the substance that transmits genetic information. From the initial discovery of the phenomenon known as "the transformation of pneumococcal types" until the identification of the transforming substance as DNA, all of the researchers were medical bacteriologists primarily interested in the cause and control of human pneumonia. Admittedly, in the latter stages of the search we came to see that our findings would not help to eradicate pneumonia, but all of the earlier steps had emerged from a study of the disease.

It is not surprising that pneumonia had preoccupied so many bacteriologists shortly after birth of the science a little over one hundred years ago. At the turn of the century, pneumonia was the leading cause of death, ranking well ahead of today's principal killers—heart disease and cancer—and it was not limited to the aged and infirm. As the precision of the techniques of bacteriological diagnosis improved, it became evident that most of this devastating pneumonia was caused by a single group of bacteria, referred to most commonly as the *pneumococci.*

Pneumococci were first isolated from human sputum after inoculation into laboratory animals in experiments reported in 1881 independently by Louis Pasteur in France and George M. Sternberg in the United States. The recognition of their relationship to lobar pneumonia came from studies in several laboratories over the next few years. The microorganism has been called by a variety of names by bacteriologists since that time, partly because of the difficulties inherent in defining the relationships of bacterial species to one another so as to permit accurate application of the Latin binomials customarily used in biology. Today they are officially known as *Streptococcus pneumoniae,* indicating their relationship to that large family of bacteria that includes those responsible for streptococcal sore throat and a number of other human ailments. I will continue to refer to them as pneumococci, both for simplicity's sake and for the historical dominance of this designation.

The pneumococcus is perhaps not especially remarkable among the vast array of bacteria in nature. It does possess certain attributes that make it recognizable to trained bacteriologists and set it apart from other microorganisms. It is about average in size, being approximately 1 micrometer (or one-millionth of a meter) in diameter. In nonmetric terminology, it would require 25,000 pneumococci lined up in a row to extend one inch. They tend to occur in pairs termed *dip*lococci, which simply means that after dividing by the usual bacterial process of binary fission the two daughter cells remain associated. This post-divisional association extends even further so that the organisms often appear as short chains, a con-

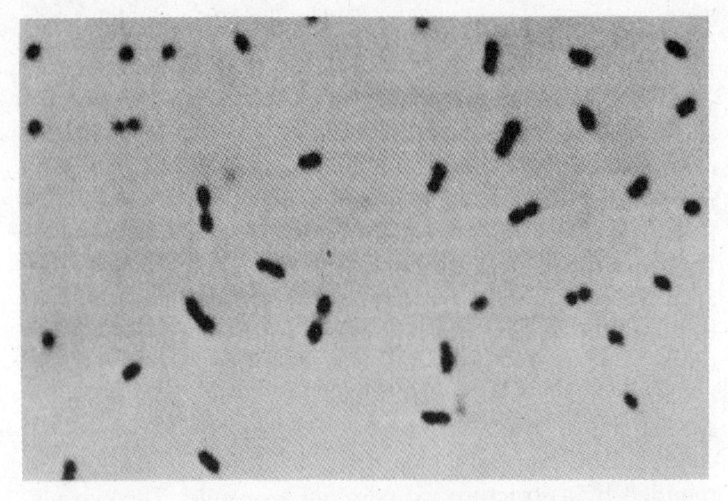

Type III pneumococci stained with gentian violet. Magnification approximately 2000×.

(Reproduced with permission from the Proceedings of the American Philosophical Society, *1984, 128:21.)*

Living type III pneumococci in the presence of India ink. Particles of ink form the dark background and reveal the capsule surrounding the organisms.

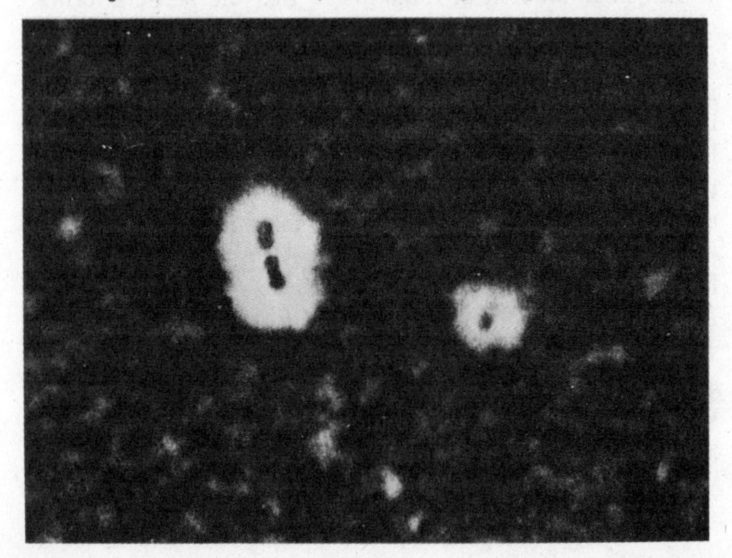

figuration typical of streptococci in general. They are frequently seen as pointed at one end and were thus referred to in the early days as "lancet-shaped." These characteristics of the organism are readily detected by examination of living bacteria with an ordinary light microscope, but special staining techniques have been applied to get additional information. Insofar as the internal structure of the organism is concerned, as in the case of other bacteria, not much was learned about this until the modern procedures of electron microscopy were used.

More relevant for us are the results of the application of special stains. These stains revealed that virulent pneumococci with the appearance just described were in reality surrounded by a structure we now call a capsule. The capsule obviously has less substance than the organism itself. Its invisibility under ordinary conditions is due to its being optically indistinguishable from the surrounding medium. Procedures other than staining can be used to confirm the presence of a capsule. In one method a suspension of organisms is mixed with a small amount of India ink so that the particles of ink clearly delineate the capsule as separate from the body of the organism. The size of the capsule varies among strains of pneumococci, but it can be very large and exceed the diameter of the coccus by three- or fourfold. As will be apparent later, the capsule is an essential feature for the virulence of the pneumococcus as an infectious agent. Furthermore, it plays a central role in the story of transformation.

For growth of pneumococci in the laboratory, a bacteriologic medium that resembles beef broth is commonly used. In fact, it is merely an extract or infusion of beef heart to which has been added a material called peptone. Peptone is a preparation of meat products that are partially broken down, usually by treatment with digestive pancreatic enzymes. Also present in the medium are some sugar and salts. After adjusting the pH (a measure of the acidity or alkalinity of the soup) to a value close to that occurring in blood, the completed medium is sterilized by one of a variety of procedures. The most common one, autoclaving, uses steam under pressure so that the temperature is well above the boiling point of water in order to ensure the killing of heat-resistant spores. When a good-quality medium of this kind is inoculated with pneumococci and incubated at about normal body temperature, the organisms will soon begin to divide, with a doubling time of 20 to 25 minutes, so that a thousand organisms will increase to several million in a matter of a few hours. A fully grown culture in this medium after 8 to 12 hours will have a popu-

Electron microscopic picture of a thin section of a short chain of type III pneumococci. The organisms were embedded in the presence of India ink before sectioning so that the capsule is made visible. Magnification approximately 25,000×.

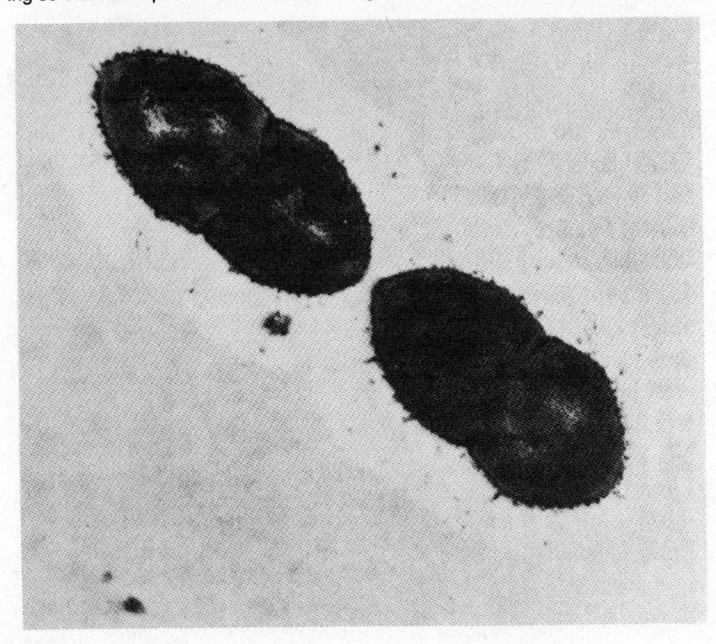

lation on the order of 500,000,000 diplococci or short chains of pneumococci per cubic centimeter* of medium. As a point of reference, it should be noted that a household teaspoon holds about 5 cubic centimeters. In dealing with these large numbers, bacteriologists have found it useful to emulate physicists and astronomers by using exponential figures. The usual notation for the number just given would be 5×10^8 organisms per cubic centimeter, or simply 5×10^8/cc. The fully grown culture has a visible turbidity imparted by the ability of the dense collection of bacterial particles to interfere with the transmission of light through the medium. However, this turbidity depends on large populations of organisms, and there can be as many as 10^7/cc without any apparent change in the clarity of the medium. The incidental information to be gained from this fact is that a solution may be very clear and still contain a great many bacteria.

The pneumococci in cultures of this kind can be harvested by centrifugation, a process that involves transferring the fluid to appropriate containers and placing them in a machine that rotates at speeds of 1000 revolutions per minute or more. The organisms by virtue of having greater density than the medium are deposited at the bottom of the container and may then be washed free of medium by suspension in salt solution, followed by recentrifugation. A mass of pneumococci that has been well packed by centrifugation is white or cream-colored with the consistency and appearance of a yeast cake, which is itself a collection of packed microorganisms. The yield is not tremendous, however, and the dry weight of pneumococci from a liter of culture (or approximately 1000 cc) ranges only from 0.25 to 0.5 gram.

An alternative way of growing pneumococci and other bacteria, one useful for isolating, identifying, and counting the number of viable organisms present in a liquid culture, is to

*The term milliliter (ml) is now generally used in place of cubic centimeter, but for our purposes the two terms are essentially equivalent.

place them on solid media. The composition of the medium is essentially the same as that of the fluid broth except that agar is dissolved in the heated mixture so that it will form a firm gel on cooling. In the case of the pneumococcus it is useful to add also a small amount of blood in order to facilitate growth of the organism on the surface of the gel. When pneumococci are incorporated in a medium of this kind or spread on its surface, each unit (diplococcus or short chain) will grow to form a colony of many millions of cells that is readily visible to the naked eye. The colonies assume a round configuration, ranging in diameter from less than 1 to 2 or 3 millimeters, and the appearance of their surface is variable from strain to strain (as well as from one species of bacterium to another), although pneumococcal colonies have certain common characteristics that are useful in identification. If a fluid culture is accurately and appropriately diluted before plating on solid medium, a reasonably reliable estimate of the number of viable organisms in the culture can be obtained by counting the number of colonies formed. For example, a culture like that referred to above with 5×10^8/cc when diluted a millionfold would contain only 5×10^2 (or 500) colony-forming units per cubic centimeter, a number within the range feasible for counting. In practice, these high dilutions of a culture are obtained by a stepwise process in which serial tenfold dilutions (e.g., 1.0 cc of culture into 9.0 cc of diluent) are prepared and designated by a negative exponential system: a 10^{-2} dilution equals a hundredfold dilution and 10^{-3} a thousandfold, etc. If 1.0 cc of a 10^{-7} dilution of a culture yields an average of 50 colonies on solid medium, it is concluded that the undiluted culture contained 5×10^8 colony-forming units per cubic centimeter. The fact that the pneumococcus tends to occur in pairs or short chains makes the figures only approximate from the point of view of the number of *individual* bacteria present and dictates the use of the term "colony-forming unit."

One of the attributes of the pneumococcus that sets it apart from most other bacteria has to do with its pronounced suicidal tendencies. The organism is endowed with a mechanism for self-destruction, composed of a set of enzymes which can dissolve its protective cell wall and which are ordinarily kept in an inactive state under favorable conditions, such as those promoting rapid growth. However, under a variety of other conditions these autolytic (self-dissolving) enzymes can be triggered into action and an entire population of the organism can be wiped out in a matter of minutes. It is not easy to see how this self-destructive tendency provides any benefits to the pneumococcus, but it was turned to the advantage of the bacteriologist long before the explanation for it was known. One of the early specialists in the field, Fred Neufeld in Germany, discovered in 1900 that the addition of a little rabbit or ox bile to a culture of pneumonococci resulted in complete clearing of the suspension after a short period of incubation. This property was called bile solubility and became widely used as one of the diagnostic characteristics of the organism. Pure bile salts and a number of modern detergents were subsequently found to be highly active in solubilizing pneumococci, presumably by triggering the autolytic system, but this result can also be brought about by nonchemical means, e.g., by repeated freezing and thawing of a suspension of pneumococci.

Another property of the pneumococcus that was exploited from the time of the earliest studies and one that proved of great value in isolating the organism from sputum or other body fluids is its striking virulence for the laboratory mouse. On injection into the abdominal cavity, the organisms multiply rapidly and usually result in the death of the mouse from a widely disseminated infection within one or two days. Pure strains of pneumococci isolated from pneumonia patients were frequently so highly virulent that a single diplococcus would cause a fatal infection, indicating that the mouse has little natural defense against these organisms when they are injected by this route. The mouse model of pneumococcal infection proved useful for many types of investigation, including studies of possible means of treating or controlling the disease.

Most of this information on the fundamental properties of pneumococci was available by the turn of the century. An increasing number of laboratories then began to seek an explanation for the extraordinary disease-producing capacity of these pathogens. Always in the picture was a possible approach to the theory and prevention of pneumonia. Nowhere was this research pursued more intensively than at the newly established Hospital of the Rockefeller Institute for Medical Research. This hospital, the first in the country to be devoted solely to the investigation of human disease, had opened its doors in 1910, a few years after the founding of the Institute. The first director of the hospital, Rufus Cole, was a Hopkins-trained physician with extensive experience in clinical bacteriology. Given this background, and the devastating impact of lobar pneumonia at this time, a natural early project in the new hospital was the study of this disease. That study was destined to continue with unabated concentration for over thirty years.

Bacteriological research had become progressively more dependent on the application of the techniques of immunology, a science that had its origins in bacteriology and had grown along with it; intensification of the immunological approach assumed a major role in pneumococcal studies. The early attempts to protect animals against bacterial infection by vaccinating them with appropriately weakened or killed preparations of the organism had led ultimately to the recognition that the sera of immunized animals contained substances that reacted specifically with the bacteria. These substances came to be known as *antibodies* and the bacterial components capable of inducing their formation as *antigens*. Antigens were clearly not limited to the bacterial world. Foreign proteins, such as the albumin of egg white, would also result in the appearance of serum antibodies after injection into an animal.

There was a remarkable degree of specificity in the antibody response. Experiments showed that the antisera directed against one bacterium had no reactivity with unrelated organisms and that antibodies to one protein, such as albumin, would not recognize other foreign proteins. There was no understanding of how animals managed to mount so specific a response to a wide variety of different antigens. This puzzle defied solution until the developments of modern immunology in the last quarter-century, but the available knowledge was applied with great vigor to the problems of infectious disease.

In the case of the pneumococcus, it was Neufeld, discoverer of bile solubility, who obtained the first solid evidence for the diversity of pneumococcal strains by using immunological techniques. Neufeld observed that sera from rabbits and horses that had been injected with pneumococci isolated from one of his pneumonia patients would protect mice from infection not only with the same strain but also with pneumococci from some of his other patients. He concluded that those strains neutralized by the serum were all alike and designated them as type I pneumococci.[1] Later he found a second strain that produced antisera against some of the remaining pneumonia cultures but not against type I strains. These were referred to as type II. Neufeld's findings by the mouse protection test could be confirmed by a test-tube test known as the agglutination reaction. In the presence of type I antiserum, type I organisms would clump together to form large masses while the cocci of other strains remained separate from one another. The results of mouse protection and agglutination experiments matched each other perfectly.

Similar efforts to sort out the varied collection of pneumococci obtained from pneumonia patients were among the first studies to be pursued in Cole's laboratory at the Rockefeller Institute. Alphonse R. Dochez, a member of the initial group of physician-scientists recruited by the research hospital, was able by Neufeld-like techniques to divide his collection into four groups. His group I and group II corresponded to Neufeld's type I and type II; group III was a smaller but nonetheless important category that had at one time been mistakenly thought to be separate from the pneumococci; and group IV represented all other strains from pneumonia cases that did not fall into one of the first three groups and for the most part appeared also to differ from one another.[2] Over the succeeding decades group IV was shown, by the same kind of serological analysis, to comprise a bewildering number of specific types, and some of the so-called "higher types"—that is, those with higher numbers than the original I, II, or III—were encountered with increasing frequency as causative agents in pneumonia. However, types I, II, and III (the designation finally accepted as a combination of Neufeld's and Dochez's) were responsible for fully three-foruths of the cases of pneumonia in 1910. Most of the intensified research of the period dealt with these organisms.

In 1913 Rufus Cole made a move that was to have a profound impact on the development of pneumonia studies at Rockefeller. During a period when he was looking for a bacteriologist at the hospital, Cole had encountered a paper by Oswald T. Avery on the subject of tuberculosis. The paper influenced him to consider Avery seriously for the job. Avery, then 35, was working in Brooklyn at the Hoagland Laboratory, a privately endowed laboratory associated with the Long Island Medical College. In later years, he liked to tell the story of Cole's visit to Hoagland to look over the unsuspecting candidate. Avery was at the laboratory bench when Cole arrived. On being asked what he was doing, Avery replied that he was testing pneumococci for bile solubility using a preparation of ox bile. Cole's comment, which Avery remembered as something of a put-down, was: "At the Rockefeller Institute we use buffered solutions of pure bile salts for this purpose."[3] My personal postscript to this story is that when I arrived at Avery's laboratory at Rockefeller twenty-eight years later, preparations of sterile ox bile were still kept on hand for testing bile solubility, although bile salts were certainly used for chemical procedures with the organism.

Cole's overall impression must have been very favorable nonetheless. He quickly followed up this visit by having Avery come to Rockefeller to meet the director, Simon Flexner, who soon thereafter arranged for Avery's appointment to start on September 1, 1913. Avery came to his new job with a varied background in bacteriology and immunology as well as an established reputation as a skilled instructor in the intricacies of his field of science, especially to small groups. He had already acquired the nickname of "Professor," a title that he never officially held. It was shortened to an affectionate "Fess" and used by colleagues, friends, and family. I learned from one of his stories, for example, that his young niece called him "Uncle Fess." Quickly immersed in the world of pneumococci, Avery quite naturally found himself associated with Dochez in collaborative studies. This led to a lifelong friendship and scientific give-and-take between the two men which was only temporarily interrupted when Dochez went to Johns Hopkins from 1919 to 1921 as associate professor of medicine. On Dochez's return to New York as professor of medicine at Columbia, they established "bachelor" quarters together, a durable arrangement which lasted until Avery left New York for Nashville in 1948. Their opportunities to stimulate one another scientifically thus did not end even when they were no longer together at Rockefeller. Many of their evenings at home involved long discussions during which they would try out their ideas on each other.

Dochez and Avery discovered in 1917 that culture fluids of pneumococci, filtered to remove the bacteria, contained a substance in solution that would form a precipitate when added to antisera of the same kind that were used for mouse protection and agglutination. This precipitation reaction proved to be type specific; that is, culture fluids in which type I pneumococci had grown would precipitate only with type I antisera, type II cultures with type II antisera, and so forth. Accordingly, they called this unidentified component of the cultures the "soluble specific substance," or SSS for short.[4] Reasoning that this substance was likely also to be released in the body from the masses of pneumocci in the lungs of pneumonia patients and then to be excreted by the kidney, they looked for it in patients' urine. Their hunch proved to be correct. It was frequently possible to diagnose the type of pneumococcus causing an infection by a simple test of the urine at the time the patient was admitted to the hospital. The diagnosis was consistently confirmed by typing the offending organism once it had been isolated. On occasion, the SSS could even be detected in the blood by the precipitation reaction. Gradually it became apparent from a variety of clues that the soluble specific substance must derive from the capsule of the organism, that tenuous halo that surrounds each virulent pneumococcus. More or less concurrently, compelling evidence had accumulated that the presence of the capsule was a *sine qua non* for the virulence of the pneumococcal cell. The most dramatic proof of this came from the finding that pneumococci, under appropriate cultural conditions, can lose their ability to form capsules. The unencapsulated strains obtained by this process, which was termed "bacterial dissociation," retained the other properties of the parent encapsulated organisms but were totally avirulent for mice. Billions of these unencapsulated cells could be injected without visibly affecting the well-being of the mouse. This contrasted to the fatal effect of a single diplococcus of the parent organism. Some notion of what was going on inside the mouse to create this remarkable result could be deduced from studies of the white blood cells, the first line of defense of the body against bacterial infection. The white cells that serve this function had been dubbed "phagocytes"—literally, cells that eat—and the process by which they engulf and destroy bacteria is called phagocytosis. Phagocytes have great difficulty ingesting encapsulated pneumococci. Something in the capsule clearly interferes with the process, and as a result the organisms can grow almost without restraint in blood or tissues of a susceptible host. On the other hand, they will rapidly engorge themselves with unencapsulated organisms under the same conditions, and a single white cell can accommodate hundreds of pneumococci without indigestion.

Experiments with phagocytosis also did much to explain the mouse protective effect of specific antisera. In the presence of type-specific antibodies, a virulent, encapsulated pneumococcus becomes fully susceptible to prompt ingestion by phagocytes. Thus, it began to appear that the antibodies that protected mice against infection, those that agglutinated the organisms, and those that precipitated with Avery and Dochez's SSS were all the same and were directed against the capsular material. Another manifestation of the interaction between specific antibodies and the capsular material was encountered by Neufeld when he observed that the usually invisible capsule could easily be seen under the microscope in the presence of type-specific antiserum. This striking effect, which he called the Quellung—or capsular swelling—phenomenon, even though it is not at all certain that "swelling" by itself is enough to make the capsules visible, was widely applied as a means of typing pneumococci in the clinical bacteriological laboratory. This was one of the first procedures that I learned in Harriet Lane in the study of sulfapyridine therapy of pneumococcal pneumonia in children.

In addition to its effects on immune reactions and virulence, the capsule also contributes a distinctive appearance to colonies of pneumococci when they are grown on the surface of agar medium. The colonies of encapsulated organisms have a smooth surface, shiny or velvety in appearance, whereas the smaller colonies of unencapsulated organisms have a rough, rather pebbly surface (see the photograph on p. 166). The terms "smooth" and "rough" derived from this colonial configuration came to be used as a shorthand equivalent for encapsulated and unencapsulated, or even for virulent and avirulent. A further abbreviation to S and R was generally accepted as the designation of these differences.

As this kind of information on the capsule and its interaction with antibodies accumulated, the rationale for the use of serum therapy in pneumonia was reinforced, and the efforts to develop this approach were redoubled in the Avery laboratory. At the same time, Avery was increasingly driven to find the chemical nature of the specific soluble substances. He was about to display two of the characteristics that were responsible for his extraordinary success as an investigator: an uncanny ability to ask the right questions and dogged persistence in finding the answers, seeking expert help whenever necessary. Some of his early observations with Dochez had suggested that SSS might be a protein, a reasonable possibility in view of the fact that only proteins were thought to act as antigens and to lead to the production of specific antibodies, but subsequent examination of the properties of the material made this uncertain. Avery continued to pursue the problem along with his other studies, making concentrated preparations of SSS which he was able to purify to some extent, but he was frustrated in his attempts to learn the chemical nature of the material. As he was looking about for help with his problem, his eye fell on Michael Heidelberger, a young organic chemist who had spent several years in the Rockefeller Institute laboratory of Dr. Walter Jacobs working on chemotherapy and had recently moved to the hospital with the group of clinical chemists under Donald D. Van Slyke. Heidelberger was intrigued by the problem when Avery presented it to him but felt obligated to complete his work of preparing crystalline hemoglobin for Van Slyke before taking on another project. From time to time, on meeting Heidelberger in the corridor, Avery would remind him anew by showing him "a small vial of brownish powder" and saying:

"When can you work on this, Michael? The whole secret of bacterial specificity is in this vial."[5] Finally, some time in 1922, they began the job together.

It was Avery's task to provide enough of the crude bacterial product to satisfy the needs of the chemist. These needs were considerably greater than anything he had been accustomed to in his own analytical studies of the material. They selected type II pneumococci for the initial efforts as the organisms with the most favorable properties. Heidelberger applied a variety of purification steps which they monitored by the precipitation reaction with specific type II antiserum. As this process went on, they obtained highly active material that was totally devoid of protein and, according to Heidelberger, Avery at one point asked, "Could it be a carbohydrate?"[6] The answer, as it turned out, was yes. The soluble specific substance belonged to a class of complex carbohydrates that are known as polysaccharides.

At this point it will probably be helpful to the reader without much biochemical background for me to include a few words about the polysaccharides. The term, which literally means "many sugars," is applied to a common class of substances that represent large molecules formed chemically by linking together the simple sugars, or *mono*saccharides. Since there are numerous known monosaccharides, each of which can be linked to others in a variey of ways, the potential for diversity among polysaccharides is very great. The most common and best-known monosaccharide in nature is called glucose, the sugar used by animal cells to provide their energy requirement and the culprit in human "sugar diabetes." When glucose is appropriately linked with another simple sugar—fructose, or fruit sugar—the product is a *di*saccharide, sucrose, the familiar household sweetener obtained from sugarcane or sugar beets. When living organisms form polysaccharides, this linking together of simple sugars is continued until up to many hundreds may be tied together in one very large molecule. Starch is a relatively simple representative of the family of polysaccharides, since it is composed solely of glucose units combined in one large structure. The polysaccharides that were identified as the specific soluble substances of pneumococci proved to be more complex, like most others found in nature, because they have two or more monosaccharides built into the structure.

The tools available to Heidelberger and Avery for the analysis of pneumococcal polysaccharides in the early 1920s did not permit them to work out the details of their composition immediately. They knew that the type II polysaccharide contained some glucose but that there were other unidentified sugars present in larger amounts. Much later it was found that the predominant sugar in the polysaccharide is one known as rhamnose, one that occurs commonly in the plant and bacterial world but not in the animal world. The situation appeared to be less complex in the case of the type III polysaccharide in which they found only two components: glucose and glucuronic acid, the latter being an acid derivative of glucose that is representative of the many modified monosaccharides that are found in nature. Together with another chemist, Walther Goebel, who joined the study in 1924, they were ultimately able to show that the polysaccharide is composed of equal amounts of the two sugars, occurring alternately along the linear molecule, so that the repeating unit is a glucose–glucuronic acid disaccharide.[7] The specificity of the type III polysaccharide is inherent in the disaccharide structure, as established elegantly by Goebel some years later when he showed that the disaccharide when coupled chemically to a protein yielded a synthetic antigen that would induce the formation of antibodies reactive with the intact polysaccharide.[8] The antibodies would even protect mice against type III infection.

Avery had achieved his goal of finding out what the specific soluble substances were made of. At the same time he had obtained the important corollary information that the protective capsule surrounding the pneumococcal cell consists primarily of a complex sugar. His private nickname for the pneumococcus, "the sugarcoated microbe," had its somewhat nostalgic roots in this heady period of discovery. The findings had broader biological implications, going well beyond application to problems of pneumococcal infection. They established that polysaccharides can express biological specificity and are able to act as antigenic substances.

Because of the revolutionary aspect of these implications, the work was not immediately accepted universally. A primary objection was the view that only proteins have the necessary diversity to display this kind of specificity and antigenicity. It was held by some that the polysaccharide preparations must be contaminated with a small amount of a highly active protein. Scientists tend to be conservative. To some extent their skepticism is justified, since many radical new findings ultimately prove to be wrong. Avery's response to this kind of skepticism was to search for some additional type of experimental evidence that would dispel the doubts about the polysaccharide nature of the specific soluble substance.

He reasoned that if he could destroy the polysaccharide by some specific means that was unlikely to affect a contaminating protein, he could confirm that the biological activity depended on the presence of the intact polysaccharide. However, the chemical procedures then available to him were not selective enough in their action for this purpose. He thus explored the possibility that there might exist enzymes that were capable of degrading the pneumococcal polysaccharides. Finding that none of the animal enzyme preparations that he tried had a detectable effect on his material, he turned to a number of plant and microbial sources—some of them rather exotic, like the papaya—but without encountering anything that altered in the slightest degree the reactivity of the polysaccharides with their specific antisera. His soluble specific substances appeared to be extraordinarily resistant to destruction by any of the enzymes that occur in the plant or animal world.

As so often happens, chance played a large role in the ultimate resolution of this problem. Sometime in the spring of 1927, a young French scientist, René Dubos, appeared on the scene. Dubos, who was about to complete his studies for a Ph.D. degree in soil microbiology at the New Jersey Agricultural Experiment Station, paid a visit to the Rockefeller Institute to see his countryman, Dr. Alexis Carrel. In the course of that visit, Dubos was introduced to Avery in the justly famous dining room of the Institute where all of the members of the scientific staff gathered for lunch. Avery asked Dubos about the work he was doing and learned that he was engaged in a study of soil microorganisms that are capable of decomposing cellulose. Since cellulose, a major constituent of the cell wall of plants, is a polysaccharide made up of glucose units, the relationship of Dubos's thesis studies to the problem that had been plaguing him was immediately apparent to Avery. He invited Dubos to join him for further discussions in his office where he related the story of the pneumococcal polysaccharides and the importance of finding an enzyme that could destroy them.

Dubos remembers Avery relating the importance of such an enzyme to its potential value in learning more about pneumococcal infections. The interpretation that I have given, relating it rather to Avery's desire to eliminate the possibility of some contaminating substance being responsible for the activity of his polysaccharide preparations, is of course secondhand and comes from having heard repeatedly in later years his recital of the events in one of his famous monologues. However, I find that this interpretation is backed up by what Avery wrote when he first reported the work on the enzyme to the Board of Scientific Directors of the Rockefeller Institute in April 1930. He talks about the action of the enzyme being "further proof that these polysaccharides, and not impurities carried along with them, are really the substances responsible for specificity." Nonetheless, I have no doubt that both motivations were behind Avery's continued interest in the enzyme, and it is also true that by the time such an enzyme was finally obtained the doubts about the polysaccharide nature of the capsular substance had been pretty well dispelled.

Dubos had indicated to Avery in that first conversation that he considered it possible to find a soil organism capable of producing the kind of enzyme he wanted. Showing again his characteristic of seeking expert help when needed, Avery made arrangements that same day for Dubos to meet both Rufus Cole and Simon Flexner. Quietly he set in motion the process of obtaining an appointment for Dubos at Rockefeller, without making it obvious to Dubos. The upshot was that Dubos became a member of his laboratory in September 1927.

On his first exposure to this new environment where the emphasis was on the study of disease-producing microorganisms, Dubos got interested in certain aspects of pneumococcal biology that were unrelated to the capsular polysaccharide and did not immediately initiate the search for a soil organism that would produce the long-sought-for enzyme. No doubt he was gently reminded of it from time to time by Avery, and he began his attack on the problem in the summer of 1928, using the type III capsular polysaccharide as the test substance. Despite the very large number of different bacteria and other microorganisms in the soil, it was soon obvious that those capable of degrading the type III antigen must be very rare. A variety of soil samples incubated with solutions of the polysaccharide for long periods of time resulted in no detectable loss in its ability to precipitate with antisera. Success finally came with the use of a sample of soil from a cranberry bog in New Jersey, from which he eventually isolated a bacterium in pure culture that possessed the required properties.

While the soil bacterium clearly made an enzyme that split the type III polysaccharide, it would only do so when the polysaccharide was present in the medium and when simple sugars that it could use as an energy source were absent. It was thus one of the early examples of an enzyme whose synthesis is "induced" by the substance it attacks. The soluble enzyme released by the bacteria could be partially purified and shown to progressively break down the polysaccharide into small units made up of only two to four monosaccharides that would no longer precipitate with specific antisera. It would also rapidly remove the capsule from living type III pneumococci and, most remarkably, cure mice that had been infected with a lethal dose of the organisms. It had no effect on the capsular polysaccharides of other types of pneumococci, being entirely specific for type III.[9]

Manifestly, this SIII enzyme, as it came to be known, had all of the attributes that Avery wanted. It could verify the polysaccharide nature of the specific soluble substance and at the same time provide a valuable tool for the study of pneumococcal infection. Dubos reports that when he wrote him of the success of these experiments in the summer of 1929, Avery took the most unusual step of returning from his customary holiday in Maine to join in further experiments with the enzyme. It is the only occasion that I know of on which he interrupted his vacation to return to New York for laboratory work, since this was a period that he liked to reserve for contemplation and renewal. It serves to underscore his deep interest in this development.

The SIII enzyme was later to be a significant factor in the analysis and purification of the pneumococcal transforming substance, but this recital of its discovery gets us slightly ahead of our story. The enzyme was not yet known at the time that Griffith carried out the first experiments on pneumococcal transformation, which he reported in early 1928. However, most of the other information on the pneumococcus so far described, including the polysaccharide nature of the capsular material, was available to Griffith. We will now change the scene temporarily to London and describe the events in his laboratory that marked the beginning of what was to become a new era in biological science.

IV

TRANSFORMATION

FRED GRIFFITH was a medical officer in the Ministry of Health in London. Working in the pathological laboratory of the Ministry in the period immediately following the First World War, he had been caught up in the same sense of urgency concerning the problem of pneumonia that had motivated other workers, like Cole, Dochez, Avery, and Neufeld. Local medical officers sent him specimens from patients with lobar pneumonia from which he would isolate and type the pneumococci. He accumulated a large number of strains of pneumococci in this way and engaged in a variety of experimental approaches in an attempt to learn more about their behavior as pathogens.

One observation that intrigued Griffith a great deal was that a single sputum sample from a pneumonia patient could harbor as many as four or five different serological types of pneumococci. His method of demonstrating this was ingenious. A sample of the sputum specimen was injected into a mouse and the organism isolated on the demise of the animal would prove to be one of the common types of pneumococci, for example, type I. He would then mix another sample of the same sputum with a little type I antiserum, thus providing protection against type I infection, and inject the mixture into a second mouse. On this occasion the animal would succumb to an infection that proved to be caused by one of the higher types of pneumococci, a member of the so-called group IV. The process could be repeated with another sample mixed with antisera against both type I and the higher type, and still another type of pneumococcus would emerge. In this manner he was able to show that many pneumonia patients harbored two or more different types of pneumococci.

Griffith found it difficult to believe that these patients had acquired each of these different bugs as a separate infection. I am not sure why he was so resistant to this idea. The dissemination of several serological types throughout the population was common at times of high incidence of pneumonia. There was thus a reasonable chance for the acquisition of two or more of them. However, his skepticism proved to be a vital element in the story of transformation by inspiring the direction of his future research. He entertained as an alternative explanation for the occurrence of multiple types in a single individual that the pneumococci in the tissues were undergoing a change in serological type as a result of immune processes or other environmental influences provided by the host. He summarized this view in the following words:

> On a balance of probabilities interchangeability of type seems a no more unlikely hypothesis than multiple infection with four or five different and unalterable serological varieties of pneumococci.[1]

His initial studies on the variability of pneumococci were directed toward a reexamination of the conditions under which the organisms would lose the ability to produce a capsule; that is, to change from smooth (S) to rough (R) with concomitant loss of virulence. He found again, as he and others had earlier, that the most consistent means of bringing about the change from S to R was to grow the S pneumococci in the presence of specific antiserum directed against the capsular polysaccharide. Similar unencapsulated R forms could be obtained by other procedures, such as prolonged incubation of cultures on the surface of special agar media. The R forms that were isolated by these procedures were all alike in that they lacked capsules and were nonvirulent for mice, but they differed greatly in the ease with which they could be reverted to S forms of the original capsular type. While some R strains seemed able to regain readily the capacity to produce capsular material and to again become virulent for mice, others were more fixed in the unencapsulated, avirulent state.

These stable R forms were of particular interest to Griffith. He explored further their potential for reversion to the fully encapsulated S form. Their lack of virulence was established by the fact that injection into a mouse of 1 cc of fully grown culture, containing the usual 5×10^8 organisms, caused no obvious ill effects. He conceived the idea of using even larger inocula by centrifuging 50–100 cc of culture and injecting the total mass of bacteria recovered in this way under the skin of a mouse. This procedure frequently gave rise to fatal infections, and in each instance S organisms of the type from which the R strain was originally derived were cultured from the mouse on autopsy. Thus, if enough of the organisms were used under the right conditions, even the stable R forms were often able to revert.

The interpretation that Griffith placed on these results led to his next experiments. He felt that such strains "may have retained in their structure a remnant of the original S antigen," and he pursued the argument as follows:

When a strain of this character is inoculated in a considerable mass under the skin, the majority of the cocci break up and the liberated S antigen may furnish a pabulum which the viable R pneumococci can utilise to build up their rudimentary S structure. The amount of S antigen in an R strain, even one only partially attenuated, might not be very large, and it might happen that such an R strain did not liberate in sufficient concentration the stimulating or nutrient substances necessary to produce reversion. It appeared possible that suitable conditions could be arranged if the mass of the culture was derived from killed virulent pneumococci, while the living culture was reduced to an amount which, unaided, was invariably ineffective. There would thus be provided a nidus and a high concentration of S antigen to serve as a stimulus or a food, as the case may be.[2]

Accordingly, Griffith tried out the effect of a mass of dead S pneumococci on the reversion of R organisms. Killing the bacteria did not present a problem, since pneumococci are quite sensitive to heat and holding them at a temperature of 60°C (140°F) for several minutes is more than enough to kill the entire population of a fully grown culture. In the first experiment that he reported, type II pneumococci were killed by "steaming at 100C" and concentrated by centrifugation so that the organisms from 50 cc of culture could be injected along with a small inoculum of living type II R organisms. Each of four mice given this mixture died within a few days of an infection caused by virulent type II S pneumococci. His notion about what was going on during the process of reversion from R to S seemed to be borne out. A control experiment carried out at the same time used steamed type I pneumococci along with living type II R organisms, and in this case the mice all survived. Thus, on the first occasion that he might have observed the transformation of pneumococcal types, nothing happened.

Griffith was well aware that steaming at 100°C was a pretty rough way to treat biological material, and in the course of a series of additional experiments he tried out a variety of other procedures for heat killing the S pneumococci. The important thing was to be sure that they were all dead. One of his favorite procedures was to heat the culture at 60°C for 2 to 3 hours, much longer than was actually needed, but insurance against the possibility that a few living organisms might remain. It was with material treated in this way that he obtained the first results that appeared to involve a change in specific capsular type. Eight mice were injected with heat-killed type I S pneumococci from 50 cc of broth culture together with a small inoculum of a live R strain derived from type II. All but two of the animals survived, but both of those that succumbed yielded type I S pneumococci when cultured. In these two mice the R cells had apparently been able to use something provided by the dead S cells to start making the capsular material characteristic of the dead cells—and to keep on doing it! How does an investigator deal with a startling observation like this? As a first reaction, he would think as Griffith must have that he had made some foolish mistake and set about to repeat the experiments with added care. In this case, the results were reproducible, even though they remained spotty with only part of the mice responding in any given experiment.

There would of course have been nothing remarkable about these findings if the heat killing of the S organisms had not been complete (99.99 percent was not enough), and Griffith focused much of his effort on this point. For example, a large number of control mice were injected with the heavy dose of heat-killed S cells without the added living R organisms, and in no instance did these animals succumb to infection or yield living S pneumococci on culture. These and other experiments, which he reported in some detail, convinced him that neither escape from the killing process nor revival of dead S organisms was a likely explanation for his remarkable findings. He carried out a number of other experiments on transformation with other strains of pneumococci. Type I R strains could be transformed into either type II or type III S, and similarly type II R strains could be transformed to type I or type III S. He also found that his various R strains differed a great deal from one another in their susceptibility to transformation just as they differed in the ease with which they would revert to the original S type. Nevertheless, change of type occurred with a high enough frequency in his repeated tests to assure him that it was a general phenomenon and not one limited to a narrow set of conditions.

Griffith wrote up these studies in a lengthy, detailed report that he submitted to the *Journal of Hygiene* on August 26, 1927. It appeared in the January 1928 issue of that scientific journal as a paper of forty-six printed pages entitled "The Significance of Pneumococcal Types."[3] The length of the paper is largely attributable to the meticulous description of the experimental procedures in each phase of his study, a reflection of Griffith's writing style that was probably exaggerated in this case by the surprising nature of the experimental results. His long discussion of these results makes it clear that his primary concern was with their implications for the epidemiology and disease patterns of pneumonia. With reference to the possible mechanism of the transformation of type, he held to a view similar to that which he advanced as the rationale for the use of a large mass of pneumococci in his initial experiments on the reversion of R to S: "When the R form of either type is furnished under suitable experimental conditions with a mass of the S form of the other type, it appears to use that antigen as a pabulum from which to build up a similar antigen and thus to develop into an S strain of that type."[4] There was no allusion to the remarkable fact that the change was permanent and that once an R form had begun to form a capsule of a new S type it continued to do so indefinitely on subculture through countless generations. Thus, something had happened to perpetuate the change. But bacteriology had developed as a science almost as if unrelated to the rest of biology. It was too early in its history to expect genetic interpretations of any phenomena encountered in the laboratory.

Other workers interested in pneumococci found Griffith's results difficult to believe in view of their experience with the stability of pneumococcal types. Their skepticism would no doubt have been greater had not Griffith been so highly respected as an investigator. However, his novel findings received confirmatory support much more promptly than is usually the case in scientific work. Fred Neufeld, whom we met earlier as one of the leading investigators of pneumococci, had visited Griffith's laboratory in London while the transformation work was in progress and had been given a preview of the essential findings. On return to his own laboratory at the Robert Koch Institute in Berlin he set about to try some transformation experiments himself. He was pretty well prepared for this, since he had recently been carrying out a series of studies on the interconvertability of R and S forms of pneumococci, with the result that in a short time he was able to repeat Griffith's findings. His paper describing his work appeared in a German immunological journal just two months after the publication of Griffith's original paper.[5]

There is a melancholy footnote to the story of these two great pioneers of pneumococcal biology, Griffith and Neufeld. Both of them became victims of World War II. Griffith was killed in an air raid during the London blitz in 1941, and Neufeld died in war-ravaged Berlin in 1945, reportedly of starvation. Griffith had some years earlier turned to a series of important investigations of hemolytic streptococci, and Neufeld had presumably had little opportunity to carry on productive experimental work after the rise of Hitler.

I of course never knew either Griffith or Neufeld and can contribute no first-hand information on either man. It is somewhat surprising, however, that Avery and Griffith never met. Neither of them seemed much interested in travel, particularly when it came to those time-consuming transatlantic voyages. As far as I am aware, Avery made only one trip abroad, in 1925, when he obtained a passport and visas for Austria, the United Kingdom, Germany, Italy, and France. I believe that he made this trip chiefly as a tourist with little in the way of laboratory visits or other scientific contacts. He did not go to Germany in 1933 when he was awarded the Paul Ehrlich Gold Medal, probably because of the illness to which I will refer later, but he submitted an address on the polysaccharide story, carefully translated into German by a colleague, to be read at the time of the award ceremony. Later he declined at least two invitations to travel to England—to receive an honorary degree from Cambridge University and the Copley Medal from the Royal Society.

When the copy of the *Journal of Hygiene* bearing Griffith's article first arrived in the library at the Rockefeller Institute, probably not until March 1928 because of the pace of transoceanic mail in those days, Avery and his colleagues were greatly interested but not entirely convinced. By a curious coincidence, one of Avery's young associates, Martin Dawson, had like Neufeld just completed a study of the interconvertability of the R and S forms of pneumococci which he had already submitted for publication in the *Journal of Experimental Medicine*.[6] He was thus similarly equipped to get on with the job of checking Griffith's results. Dawson was very thorough in his analysis, raising a number of new points in addition to the old one of how dead the heat-killed cells were, but in the end his results confirmed Griffith in every detail.[7] A suitable R strain appeared to be convertible to any S type he chose as long as the conditions of the mouse experiment followed those described by Griffith.

Griffith had reported unsuccessful attempts to simplify the transformation experiment by carrying out the procedure in a test tube, thus eliminating the mouse as a silent partner in the process. In his initial paper, Dawson described similar attempts which proved to be equally futile. However, he remained convinced that transformation in vitro—that is, outside the living body of the mouse—should be possible and that achieving it would be an important step toward better understanding the phenomenon. He acted on this conviction after leaving the Rockefeller Institute to assume a position at the College of Physicians and Surgeons of Columbia University where he resumed experimental work on the problem with the assistance of a young Chinese associate, Richard Sia. They tried, using test tubes, a variety of manipulations of the conditions under which the suspension of heat-killed S organisms (Dawson called this a vaccine for short) were brought together with living R cells, including the addition of a number of substances like serum that were meant to help simulate the conditions in the mouse. The key to success turned out to be very simple, as is so often the case in research of this kind. It depended on controlling the numbers of living R cells that were used as an inoculum. If they used a small volume of an undiluted culture of R pneumococci, which still contained many millions of organisms, the results were always negative. However, when the inoculum contained only a few thousand organisms, as when a drop of a 10^{-4} dilution of the culture was used, they began to see the emergence of transformed S organisms almost as consistently as in the mouse.[8]

All that was needed was a little bacteriological medium suitable for the growth of pneumococci, a dash of anti-R serum (that is, a serum containing antibodies that agglutinated R pneumococci), the heat-killed type III vaccine, and a small number of living type II R organisms. After 48 hours of incubation, living type III organisms could often be isolated from the culture. After they had devised a means of obtaining positive results, they were surprised to find that they didn't really need the large amounts of killed S cells that were apparently necessary for Griffith's "nidus" in the mouse experiments. Killed organisms from as little as 0.1 cc of culture were sufficient on occasion to bring about transformation in the test tube.

This last observation encouraged Sia and Dawson to see whether they could do without the intact heat-killed cells. They tried to replace them with such things as the supernatant fluid remaining after removing the killed cells by centrifugation or with purified preparations of specific polysaccharide, which they obtained from Heidelberger who was by this time also at Columbia. When these attempts failed, they turned to extracts of living S organisms prepared by repeated freezing and thawing—as many as twenty-three times!—a heavy suspension of the bugs. As pointed out in the last chapter, this procedure triggers the autolytic enzymes of pneumococci and leads to their complete disruption, but Sia and Dawson could not know that among these enzymes were also ones that destroy the active substance they were after. They were unable to report the transformation of pneumococci in the absence of a whole-cell vaccine. In discussing the results of these experiments, however, they made some interesting comments that presaged some of the views about transformation that were to be expressed at a later date:

In considering the nature of the mechanisms by which transformation of type is effected two possibilities present themselves: either a latent attribute of the R cell may be stimulated by its association with the S vaccine, or the organisms may acquire a new property from the vaccine. The former conception involves the assumption that all pneumococci possess the latent capacity of elaborating any one of the known varieties of specific polysaccharide associated with S organisms. The latter hypothesis suggests the possibility that, at times, certain attributes of bacteria may be transferred from organisms of one type to those of another type of the same species.[9]

I was surprised to discover recently that still another member of the Avery laboratory was involved in work on transformation during this early period. This was Thomas Francis, Jr., who was a member of the laboratory from 1928 to 1936 and whom I got to know quite well in later years. However, I cannot recall that either he or Avery ever mentioned his participation. This came to light as a result of a talk that Francis gave at the First International Congress for Virology in Helsinki, Finland, in 1968. His talk, which he entitled "Moments in Medical Virology," included the following section:

The second Moment of this revolution relates to the transformation of pneumococcus. Just 40 years ago this month, I reported to the Hospital of the Rockefeller Institute for Medical Research to work on the pneumonia service under Cole as clinician and under Avery as investigator. . . . Few of the staff were around and I had no patients. Henry Dawson was there, however, working enthusiastically on the transformation of pneumococcal types. I had read Griffith's reports and knew a little about bacterial variation—particularly the rough–smooth alteration. So I spent the mornings in the laboratory learning of these phenomena and the afternoons in the library and on the tennis court developing a model of the double fault. Being convinced that the induced change of pneumococcus types in the animal host was a true bill, I began very primitive efforts to obtain transformation in the test tube. . . . But it seemed likely that whatever the transforming principle was, it needed special care and I began making extracts by freezing and thawing the organisms in the cold under relatively anaerobic conditions so as to avoid an enzymatic destruction of the principle. . . .

New lines of effort were freely allowed even if they were not always enthusiastically supported. I found this when I studied transformation of the rough Type III to virulent in rabbits; there was a lot of specificity involved and much work, but it was never published until later (by others).[10] We called this Fess's pocket veto. In any event, Alloway came to work with me in the clinical field and I turned over to him the *in vitro* studies—which he successfully pursued. In the meantime Dawson and Sia had also succeeded. . . .[11]

I am not sure about the total accuracy of Francis's recollections, but there is no doubt that he worked actively in the field for a time. The Alloway to whom he refers was J. Lionel Alloway, who arrived in the lab the year after Dawson's departure. He knew of Dawson and Sia's success with transformation in the test tube from a preliminary report published in 1930 and also possibly by word of mouth, since Dawson maintained fairly regular contact with Avery for some years after he left his laboratory. Armed with this information about the experimental conditions that permitted transformation in vitro, he concentrated on the next step of trying to prepare active cell-free extracts of S pneumococci. His first paper on the subject,[12] which appeared in the *Journal of Experimental Medicine* in January 1932, didn't tell anything of the false starts and frustrations that he encountered in achieving this goal but merely provided a brief description of a method that seemed to work. Like Dawson and Sia, he used the procedure of freezing and thawing but carried it out rapidly and repeated it only as many times as was necessary to break up all of the organisms—not more then seven or eight. He than promptly heated the suspension of disrupted cells at 60°C for 30 minutes, followed by centrifugation to remove insoluble material. His next step was an important one theoretically as a final answer to those skeptics who continued to harbor a suspicion that the supposed transformation represented nothing more than the survival of an occasional viable S organism in the vaccines. He diluted his extract and passed it through a bacterial filter of a type made of porous porcelain (called a Berkefeld filter after its inventor) that had been shown to reliably hold back bacteria while allowing the passage of soluble substances. The filtered extract was concentrated about tenfold and tested in the transformation system.

A filtered extract of this kind prepared from type III S organisms, when mixed with some pneumococcal broth and a little anti-R serum and inoculated with a drop of diluted culture of a type II R strain, after incubation for 24 hours yielded encapsulated, virulent type III pneumococci. With the same type II R strain, a type I extract produced type I S organisms, although somewhat less consistently. There was little doubt, therefore, that Alloway had succeeded in getting the substance responsible for the transformation of pneumococcal types into solution. Thus, in the historical development of the discovery, Dawson gets credit for bringing the Griffith phenomenon out of the mouse and into the test tube, and Alloway for eliminating the need for intact, heat-killed cells by preparing active extracts. Dawson was obviously close to the latter result also. In fact, Alloway notes in his first paper that Dawson had told him personally of obtaining transformation with extracts that had been exposed to fewer cycles of freezing and thawing, but he apparently chose not to pursue this further or to publish his findings because of Alloway's priority.

Alloway, however, was not fully satisfied with the procedures he had developed. In discussing his results, he gave some hint of the difficulties he had encountered. Not all of his extracts were active by any means, and many of the successful ones were so weak as to give positive results only part of the time. This led him to continue the search for a better way of preparing active extracts, and he reported his progress in these efforts in a paper that was not completed until after he had left the Rockefeller Institute and was published in February 1933.[13] He described a number of improvements, the most important being the preparation of the extract by lysing the organisms with the bile salt sodium deoxycholate, rather than by repeated freezing and thawing. In retrospect it is surprising that neither he nor Dawson had tried this earlier, since as noted in the preceding chapter the so-called bile solubility of pneumococci had long been known. When a heavy suspension of pneumococci, with a creamy consistency, is treated with a small amount of bile salt it quickly becomes a thick, viscous mess in which intact organisms can no longer be found on microscopic examination. In an attempt to minimize the loss of transforming activity during this lytic process, Alloway began the procedure in an ice bath and after 10 minutes brought the suspension slowly to 60°C to stop the reaction and inactivate the autolytic enzymes. The whole operation could be carried out much more rapidly than the more cumbersome process of freezing and thawing.

In the subsequent handling of the extract, Alloway then introduced another new procedure that became an indispensable part of all work on the transforming substance from that time forward. He added pure alcohol in a volume five times that of the extract which resulted in precipitation of most of the material that had been released from the pneumococci and left behind the bile salt. The precipitated material could be redissolved in salt solution and shown to contain the active substance in transformation tests. Alcohol precipitation and resolution could be repeated at will without loss of activity. Alloway described the solutions of active material obtained by these procedures as slightly turbid and opalescent, but he was able to convert them into water-clear extracts by treatment with powdered wood charcoal.

Another Alloway innovation became a part of the routine procedure of the transformation studies for many years to come. He found that certain pathological fluids from human patients, usually chest fluids, were more consistently effective in supporting transformation than the rabbit or swine anti-R sera that had been used previously. These fluids also contained antibodies that agglutinated R pneumococci but appeared to have other attributes that made them superior to the animal sera. As material removed from patients for therapeutic purposes, which would ordinarily simply be discarded, they had the additional advantage of often being available in large quantities. Hundreds of fluids of this kind were screened for their effectiveness over the next decade in the Avery laboratory.

This description of Alloway's success in obtaining transformation with filtered, water-clear extracts of S pneumococci suggests that by the time his work was completed and published in 1933 all the necessary preparations had been made to proceed with the business of identifying the active substance in the extracts responsible for transformation. It was unfortunately not quite that simple. Even though the experiments had demonstrated clearly that a soluble factor was involved, transformation in the test tube remained an inconstant and maddeningly unreliable process. Sometimes it worked and sometimes it didn't, and each time it failed it was necessary to stop and try to find out why. The problem, as we shall see later, could be with any one of the components of the system so that correcting it was not necessarily easy. In short, the transformation system had not yet been developed to a point where it could serve as a dependable test in monitoring the fractionation of extracts and purification of the active substance. There was obviously much work to be done before one could hope to make significant progress toward the goal of determining the nature of what had come to be known in the Avery laboratory as the "transforming principle" and abbreviated as T.P.

There is no doubt in my mind that Avery had already set his sights on this goal while Alloway's studies were in progress. His name did not appear on the papers that either Dawson or Alloway published on the work carried out in his laboratory, but he was almost certainly intimately involved and closely followed the development of the experiments. A case in point is the passage of the extracts through a Berkefeld filter, a step that was looked upon as the final proof that intact S organisms could not provide the source of the apparent type transformation. In his recital of the early history of the subject, which I first heard on my arrival at Rockefeller several years later, Avery implied that it was at his insistence that this additional evidence was sought. When Alloway's first attempts failed, with the loss of all activity after filtration, it was Avery who suggested the successful device of making the extract somewhat alkaline before passing it through the filter. As a result of previous experience, he had known that many substances would stick to the filter if the solutions were even slightly acid.

However, there was a matter relating to Avery's health during this period that certainly must have curtailed his personal participation in the research and may have even prevented him from keeping as close a watch on it as he would have liked to. He had developed the symptoms of hyperthyroidism, or Grave's disease, and suffered from gradually increasing disability brought about by this disorder. I do not know the date of onset of these symptoms, but among the papers of Simon Flexner there is a letter to Avery dated October 6, 1931, thanking him for a summary of his work and including the following paragraph: "I am sorry to have imposed the task on you, but I hope it did not cost you too much effort. And let me add that I hope you will take proper care of yourself, beginning immediately."[14] His illness had apparently progressed far enough by this time for it to have come to the attention of the director. In addition to the physical disability caused by the ailment, it also resulted in some mental anguish and depression. Not the least of his difficulties came from the tremor associated with hyperthyroidism which interfered with his almost compulsively meticulous bacteriological technique. His inability to carry out the customary bacteriological procedures without the risk of contamination was by itself a blow to his self-confidence in the laboratory.

Although the precise date on which Avery finally came to surgery for his disorder is not known, since the hospital records have all been destroyed, the available evidence suggests that it was in 1934. This was the only year of his long tenure at the Rockefeller Institute for which the report submitted each April to the Board of Scientific Directors did not include a section written by Avery. His recovery to full health following thyroid surgery was apparently slow, and it was more than a year afterward that he regained his normal weight, which was not much over a hundred pounds in the first place. There was thus a long period in the early thirties, a crucial period in the development of the research on transformation, when Avery's work in his laboratory must have been much less intensive than he would have liked.

The work on transformation had nonetheless continued. On Alloway's departure in the summer of 1932 another young physician, Edward S. Rogers, joined the pneumonia service for laboratory training and decided to take up the struggle where Alloway had left off. He had no great luck with the project during his two years in the laboratory, turning up no new information that was considered worthy of publication. I do not have the laboratory notes from this period or from the earlier work of Alloway, but the annual report to the Board of Scientific Directors in April 1933 included a summary of Rogers's investigations up to that time. He worked with extracts of type III pneumococci prepared by the Alloway procedure in attempts to purify further the active substance. None of the several procedures that he tried during that first year had any special promise. There are very few hints as to the nature of his subsequent efforts, since the following year was, as just noted, the one in which no general report from the Avery laboratory was submitted. The frustrations that Rogers felt were but a harbinger of things to come in the search for the identity of the transforming principle, a search that was taken up immediately after his departure by a new recruit, Colin MacLeod.

V

ENTER MacLEOD

COLIN MUNRO MACLEOD had skipped so many grades during his early schooling in Canada that he was only 15 years old when he first presented himself for admission to McGill University. Forced by school regulations to defer his start for another year until he was 16, he was still able to complete both premedical and medical studies by the relatively early age of 23. After two years of residency training at the Montreal General Hospital, he joined the staff of the Hospital of the Rockefeller Institute in the summer of 1934 as an assistant resident physician and assistant in medicine, attached to the pneumonia service of Cole and Avery.

MacLeod appears to have had little prior research experience, although I have no first-hand information on this point and must rely on indirect evidence. There are no items in his bibliography dating back to the pre-Rockefeller period. Nor do I have any idea what influences directed his interests to research and brought him to Rockefeller. These are among the many questions that it never occurred to me to ask Colin while I still had the opportunity. Another such question relates to how he coped with that period of his initial weeks in the laboratory when his principal concern must have been the selection of a research project. This could be a difficult process, since Avery never *assigned* a project to anyone, and after some background reading and listening to Avery's discourses on the past and present activities of the laboratory, the neophyte was expected to come forward with his own research ideas within the context of pneumonia. Whatever the situation may have been, MacLeod did not waste any time getting down to work on the pneumococcus as demonstrated by his laboratory notes, which date back to early August 1934 when Avery was almost certainly away for the summer.

My collection of MacLeod's laboratory notes is not complete, but the notes I do have give a good picture of his early research activity at Rockefeller. They are nearly all carefully dated, which is not always the case with the records of scientists' experiments, and only occasionally did I have to guess at the year in which a given set of data was obtained. It is fortunate that he was so precise, since the notes are not bound together but are recorded on loose sheets that he had organized by subject and collected in manila folders. The notes have the further happy quality of having been written in a legible hand so that deciphering was never required.

MacLeod began by becoming intimately acquainted with the pneumococcus in a series of experiments that were pretty much a repetition of those of Griffith and Dawson on the conversion of the S to the R form by growth in the presence of type-specific antiserum. It is not evident that he was doing this with the intention of finding an R strain that was more suitable for transformation experiments, but this is the way it turned out. In the course of a series of trials that involved pneumococcal strains of different types, he isolated a rough variant of a type II pneumococcus that had originally been obtained in its virulent form by Avery from a fatal case of pneumonia on the last day of 1916. MacLeod had selected this particular R variant after thrity-six serial subcultures of the parent type II strain in the presence of specific type II antiserum, and it thus came to be designated as strain R36. Because of its favorable properties, descendants of R36 became the organism of choice for all subsequent transformation experiments and are still used today throughout the world for this purpose.

The first recorded use of R36 by MacLeod in a transformation experiment—and one of his earliest attempts to carry out transformation—occurred on October 19, 1934. He applied the Dawson technique with a heat-killed type III vaccine as the source of the transforming agent. In his comments on the satisfactory quality of R36, he concluded that "transformation occurred in 100% of cases. No homologous reversion occurred." This later property proved to be characteristic of the strain. It shows no tendency to revert to the production of the original type II capsular material even when injected into mice in large quantities by the Griffith technique. Only when given type II transforming principle does it resume its original character. This stable characteristic, together with the relative ease with which it could be transformed to a number of other specific types, set R36 apart from most of the other R strains that had been isolated in various laboratories.

He soon sought to make extracts by Alloway's procedure, so that by November 22 he was able to record successful transformation of the same R strain to type I with a filtered extract of type I pneumococci. In the course of the next few weeks, he found that type III extracts gave more consistently positive results than those of type I, which led him to focus most of his attention on the former. This marked the beginning of the application of a single model—that of type II R → type III S—in the future analysis of transformation, an important step for a number of reasons, not the least of which was the availability of the Dubos SIII enzyme as an aid in sorting out the part played by the capsular polysaccharide, which was inevitably present in large amounts in the extracts.

Throughout this period there are no notes in Avery's hand or other evidence of his personal participation in the experimental work. However, a comment by MacLeod in the notes describing his third type III extract, prepared from 4.8 liters of pneumococcal culture on January 4, 1935, indicates that Avery was almost certainly beginning to get involved. After a brief description of the procedures used in obtaining the extract and a note on its potency, MacLeod added the following item: "*Disposition:* 125 cc to Dr. Avery." His active participation is also confirmed by the tone of the report that was submitted that spring (April 20, 1935) to the Board of Scientific Directors. The lead item in the report from the Avery laboratory dealt with transformation and, following an extensive review of the earlier work on the problem, the renewed efforts were introduced by this statement: "The work begun by Dr. Alloway and continued last year by Dr. Rogers is being actively carried on at present in collaboration with Dr. MacLeod in an attempt to ascertain the nature and properties of the transforming principle present in active extracts of S cells." Most of the properties of the transforming principle discussed in the report came from observations that represented extensions or refinements of the work of Alloway and Rogers and they were not entirely new. There was, however, one finding of major significance for future progress of the research in the direct demonstration that the autolytic enzymes of the pneumococcus could indeed readily destroy the activity of otherwise quite stable transforming extracts. This not only provided an explanation for the great variability experienced in obtaining active extracts but also suggested a possible experimental approach to improving the situation by inhibition of the offending enzyme.

This 1935 report also contains some interesting discussion indicating that Avery and MacLeod had yet to consider transformation as a possible transfer of genetic information. They wrote that "the results indicate that all R cells . . . possess a potential but inactive system of enzymes capable of synthesizing any of the type-specific polysaccharides, the particular one being produced being determined by the specificity of the activating stimulus. Once the capsular function has been specifically activated, the newly transformed cells continue to synthesize the same capsular material and retain the same type-specificity through innumerable transfers on artificial media without the further addition of the activating substance initially employed to induce transformation." In other words, they favored the first alternative mechanism proposed by Sia and Dawson. Given the large number of pneumococcal types, each with a chemically different capsular polysaccharide, this concept would seem to burden the pneumococcal cell with an inordinate amount of generally useless enzymatic material. Nonetheless, this represented a more conservative view of the situation at that time than the notion of genetic transfer, and they were still left with the remarkable observation that the "activating" transforming substance must continue to be duplicated indefinitely by the transformed cells. This was difficult to explain in the light of the knowledge then available about cellular mechanisms.

A final paragraph in this report reveals that the clinical problem of pneumonia continued to be a motivating factor in the work on transformation. They wrote that "the thought suggests itself that were we in possession of knowledge pertaining to the nature of the substances which serve as activators and inhibitors of the capsule-producing enzymes, the knowledge gained might afford a new approach to a specific attack directed toward the suppression of the capsular function upon the activity of which the pathogenicity of Pneumococcus depends." In other words, since only encapsulated organisms are able to cause disease, an understanding of how capsule production is controlled might lead to a mode of therapy based on inhibition of this process.

Despite some obvious gaps in the laboratory notes, it is evident that MacLeod's work over the next two years dealt with various aspects of the problem. Progress was infuriatingly slow, however, and they elected to include nothing on the subject in the next report in 1936. He defined a number of new properties of the transforming extracts, summarized in the 1937 report, that served to guide future investigations.

He found, for example, that transforming activity was highly susceptible to inactivation by ultraviolet light, but the full significance of this observation was not appreciated for some time. A more immediate payoff came from his application to the extracts of a procedure for removing proteins from solutions of polysaccharides that had been devised by M. G. Sevag, a worker in Neufeld's laboratory in Berlin.[1] This involved shaking the solution with chloroform together with a higher alcohol to prevent foaming, followed by centrifugation of the mixture, after which a solid cake of denatured protein would appear at the boundary between the heavy chloroform layer and the lighter aqueous extract. By removing the extract and resubmitting it several times to the chloroform process, preparations could be obtained that contained no detectable protein by the available qualitative tests but which had lost none of their transforming activity. This supported the notion that the transforming principle was not protein in nature as had been suggested earlier by the finding that trypsin, a protein-splitting enzyme from mammalian pancreas that had recently been obtained in crystalline form by Northrop at the Princeton branch of the Rockefeller Institute, did not affect the activity of transforming extracts. "Sevaging," as the chloroform procedure came to be known in the laboratory, was from then on a part of the purification process applied to all transforming extracts. Over the next decade large volumes of choloroform and many hours of effort were devoted to it.

Another line of investigation that was prominent in the shaping of future studies dealt with sources of enzymes other than the pneumococcus itself that appeared to destroy the transforming substance. Certain crude enzymes from mammalian sources, notably one that was designated as "bone phosphatase," would consistently eliminate all transforming activity after a brief exposure to the extracts. Perhaps of greater immediate significance for the research was the finding that all mammalian sera, including the chest fluids that were being used in the culture medium in which the transformation experiments were carried out, contain small amounts of a similar enzyme. Therein lay a clue to at least one of the uncontrolled factors that was responsible for the exasperating variability in the success of transforming tests. MacLeod found that heating the chest fluids at 60°C for 30 minutes seemed to inactivate this enzyme and at the same time increased their efficacy in the transforming system. It became routine to treat the chest fluids in this manner in all future studies. He noted that the enzyme in certain materials, such as rabbit serum and pneumococcal autolysates, was not inactivated at this temperature and required heating to 65°C for 30 minutes.

As a matter of fact, MacLeod had begun in 1936 to concentrate the lion's share of his attention on the serum factor required in the transformation reaction. In the 1937 report, he ended his brief comments on the properties of transforming extracts with a statement that no selective means of isolating the active fraction had been discovered and then moved on to a discussion of the "accessory factor present in serum," which he said had been more extensively studied. Most of the remainder of the report deals with findings just described on the presence in serum of an enzyme capable of inactivating the transforming principle rather than with the positive attributes of serum that made it essential in the transforming system. On the other hand, the laboratory notes show that he had carried out a large number of experiments applying a variety of fractionation procedures to serous fluids in an attempt to identify the "accessory factor." These experiments generally gave inconstant or nonreproducible results that proved to be largely uninterpretable. In retrospect it would appear that the procedures used in chemical fractionation of the serous fluids tended to introduce modifications in the material that interfered in one way or another with the complex process of transformation. Years later, toward the end of my period in Avery's laboratory, I wasted a good deal of time and effort on similar experiments directed toward identification of the serum factor. The problem was solved shortly thereafter by Rollin

Hotchkiss, who joined the laboratory after my departure and brought a fresh outlook and new approach to its solution. The answer was embarrassingly simple. Serum albumin, the most abundant single protein in serum, when added to culture media in purified form, along with a little anti-R antibody, was able to support transformation. Presumably it acted by virtue of an ability to combine with and neutralize certain substances that were toxic to pneumococci. However, since this refinement came rather late in the game, all of the tests during the identification of the transforming principle were done with medium containing heated chest fluids.

An assessment of the situation in the summer of 1937 suggests that MacLeod and Avery were making some progress and that they had succeeded in defining the conditions required for transformation in the test tube more satisfactorily than had previously been the case. To be sure, there were no breakthroughs in the search for the identity of the transforming principle, their major goal, but one could hardly consider the prospects for success in this venture hopeless. How can one explain, then, that the project was suddenly dropped at this point, with little evidence of activity over the ensuing three years? The topic of transformation was not referred to in the spring reports of 1938, 1939, and 1940. The surviving laboratory notes contain no items that were dated during this entire period, except for the records of one or two sporadic experiments. It really looks as though the whole matter was put on the shelf.

Some have interpreted this silent interval as evidence that Avery and MacLeod were not sufficiently aware of the biological importance of the problem. I believe the answer is considerably more complex. Colin MacLeod had been working for three years at Rockefeller and had as yet published almost nothing relating to his experimental laboratory studies. This would be disastrous for an aspiring young investigator today. Even in the 1930s it was a threat to MacLeod's career development. It is instructive to compare his early record with that of his friend and colleague, Frank L. Horsfall, Jr., who had also come to the Avery laboratory from McGill University in 1934. By the end of 1937 Horsfall had more than a dozen full-length scientific papers to his credit, most of them having appeared in the *Journal of Experimental Medicine.* During the same period, MacLeod's name had appeared as co-author (with Horsfall, among others) on a few papers describing clinical studies, chiefly on the use of rabbit anti-pneumococcal serum in the treatment of pneumonia, and he had a preliminary note with Dubos on a topic unrelated to transformation. The contrast was sure to be obvious to his superiors, especially to the new director of the hospital, Dr. Thomas M. Rivers, who had succeeded Rufus Cole in July 1937. Avery himself must have been sympathetic with MacLeod's predicament, since he was well aware of the difficulties of research on transformation and had a personal interest in seeing it go forward.

Colin did try his hand at a paper on the transforming substance, probably in mid-1936, and indicated that he was planning a second one on the serum factor. This information comes from a draft of the manuscript, handwritten in pencil, that was found among his papers after his death. It was not complete but certainly came close, since it included all of the experimental part along with several tables. Avery had gone over the draft, and there are pencilled comments by him sprinkled throughout the twenty-one pages. At one point, for example, in a section on the activity of the extracts, MacLeod had written: "Thus an extract was considered of low potency if 0.1 cc did not effect transformation," and Avery made the following suggestion: "Define in positive terms—An ext considered potent if 0.1 cc did effect. . . ." In the end, their enthusiasm for this interim publication must have flagged because they were still so far from the ultimate goal. There is no evidence that it got beyond this initial stage.

There was an additional factor influencing a change in the direction of MacLeod's research other than the need to find a more immediately productive line of endeavor. The laboratory had gradually worked out the details for the preparation and use of type-specific rabbit pneumococcal antisera in the treatment of pneumonia, and the accumulated experience with this approach showed that it was highly effective in reducing mortality from the disease. At this juncture, the sudden appearance on the scene of the sulfonamide drugs as potential chemotherapeutic agents in pneumonia changed the picture. If these drugs lived up to their preliminary promise, they would have the great advantage of acting on all pneumococci, regardless of type, thus eliminating the need for typing the offending organism before starting treatment. They would also obviate one of the major disadvantages of serum therapy, the frequent development of serum sickness from an immunological reaction to foreign serum. It was clearly important for the laboratory to have a look at the sulfonamides and to form its own judgment about their potential. This is the task that MacLeod undertook.

One might still ask whether it was really necessary to give up the transformation studies so completely and why, for example, Avery did not pursue these experiments on his own after MacLeod had turned to other things. In point of fact, Avery had never returned to the pattern of initiating his own laboratory experiments after his illness. He loved to discuss and participate in the devising of experimental protocols but only rarely would he then proceed to the mechanics of setting up the experiment. He would enthusiastically work with his collaborators in this effort but would not initiate it on his own. Even his nagging curiosity about the nature of the transforming substance was not strong enough to break this pattern. He no doubt followed closely MacLeod's successful experiments on the sulfonamides, and together they even returned to some experiments on another problem that had been set aside earlier. The latter relates to a discovery that Tillett had made when he was in the laboratory in 1930 on the appearance in the blood of patients with acute infections of a substance not present in normal subjects. Because of its implied relationship to the host response to infection, this substance in Avery's view clearly required better definition. Avery and MacLeod showed that it was a protein, which they obtained in pure form and characterized in some detail. This protein is still known today as "C-reactive protein," denoting its original discovery through its reactivity with fraction C (or the somatic C polysaccharide) of the pneumococcus, and studies of its function and significance are still in progress in many laboratories. It may well emerge as the third major contribution of the Avery studies on pneumonia to unrelated areas of biology and medicine, along with the discovery of the specificity of the polysaccharides and the uncovering of the genetic role of DNA.

This hiatus in the experiments on transformation raises still another question. Weren't Avery and MacLeod afraid that someone else would steal a march on them and seize the initiative in carrying out the search for the nature of the transforming substance? As unbelievable as it may seem in today's competitive scientific world, they had nothing to worry about on this score, whether they knew it or not. All of the earlier work, including that relating to transformation with cell-free extracts, had been published for some time, but no one else seemed motivated to pursue the transforming principle. The full explanation for this can never be known, but Avery's laboratory remained the only place that this line of research was pursued. If any other investigators shared Avery's intuitive guess about the significance of identifying the transforming substance, they appeared not to have the will, or perhaps the expertise, to become involved in so recalcitrant a research project. It has been pointed out, in addition, that Rockefeller was one of the few places where one could have continued financial support for such a project for so long a period without more substantial returns. In any event, except for the publications from Avery's laboratory, it was difficult even to find any mention of the desirability of learning the nature of the active substance.

By the summer of 1940, MacLeod's productive work on

the sulfonamides and C-reactive protein was well on the way to repairing any deficiency he may have had in scientific publications. Having not for a moment lost their interest in the transformation problem, he and Avery must have chafed at the circumstances that had kept them from getting on with it. They had no doubt discussed it frequently during this long interval. It is my impression—perhaps gained from something Avery told me later—that at this point in 1940 they reassessed the situation and agreed that after the summer holidays they would join forces in a renewed attack on transformation, making it their principal laboratory project. It was thus that the sustained effort that led ultimately to the identification of the transforming principle began in the fall of that year.

VI

THE RENEWED ATTACK

THE PRIMARY GOAL of this rekindled activity, as Avery saw it, can be simply stated. Studiously avoiding preconceived notions, he wanted to determine what substance in the active extracts was responsible for transformation—or, at least, as he put it, "to what class of substances it belonged." The difficulty lay in the great complexity of the extracts that were made by dissolving the pneumococci with the bile salt. Presumably, the extracts contained in solution essentially everything that had been present in the living bacterial cell, and the tools so far developed by biochemists for separating individual components of this kind of mixture were limited. They were tackling a tough problem.

It was October by the time MacLeod and Avery got down to business on their proposed collaboration. Even though I later became well acquainted with their laboratory setting and got to know both of them intimately, I have difficulty visualizing just what kind of working arrangements they were able to devise, since they were quite different in their experimental styles: Avery precise and methodical and MacLeod much more impulsive and impatient. However, the record confirms that they were cooperating closely on the project. While most of the laboratory notes were kept by MacLeod, there are frequent entries by Avery, often with both of them participating in the write-up of a single experiment. The notes for this year, extending into the summer of 1941, are all together in a single loose-leaf binder and in some semblance of chronological order. There are no significant omissions.

The first page of this notebook gives the description of an experiment dated October 22, 1940, and written entirely in MacLeod's hand except that at the top of the page, above the title of the experiment, Avery had written "Exp. I (T.P.)." This is a clear indication that Avery looked upon the resumption of research on the nature of the transforming principle (or T.P.) as a new beginning, but he was not wholly accurate in designating this as "Exp. I." A little further into the book there are two pages out of sequence indicating that a week earlier MacLeod had prepared the first type III transforming extract of the new series which he had labeled ExI / 40. The focus of these initial experiments came from MacLeod's earlier demonstration that the autolytic enzymes of the pneumococcus could destroy the transforming substance and his conclusion that it would be well to find some inhibitor of the offending enzyme so that the active material could be protected during the extraction process. In the sole example I have found of notes relating to work on this project during the previous year, MacLeod had tested several substances as inhibitors and found that sodium fluoride seemed to exert some preventive effect on the inactivation of an extract by added autolytic enzymes. Stimulated by this finding, which had been set aside for many months, they returned to reexamination of the fluoride effect and its possible usefulness in the preparation of transforming extracts.

The first extract prepared with the addition of sodium fluoride during the process of dissolution of the type III pneumococci with deoxycholate was active, although not notably more active than the extract prepared the previous week without fluoride. Nevertheless, they were stimulated to use this procedure in their first large-scale preparation made from 36 liters of a culture of type III pneumococci. This was an innovation, since earlier extracts had always come from relatively small batches of culture (3 to 5 liters), and considerably more effort was required in dealing with the larger volumes. The results were a great disappointment. The extract showed no transforming activity whatever on repeated test, and this particular approach appears not to have been pursued further even though fluoride was still looked upon as a potentially useful inhibitor.

The introduction of the use of large-scale cultures in the project requires further comment because of its importance in the research and because it provides another indication that MacLeod had been giving some attention to the problem before the renewed effort began. As noted previously, the mass of bacteria obtained from a liter of culture is not very great—not more than a few tenths of a gram—so that the amount of material available for extraction limited the progress of the experiments. Clearly, the attempts to fractionate the extracts and purify the active material would be aided by a larger mass of the packed bacterial cells. The difficulty lay in recovering pneumococci from the broth culture by centrifugation. The ordinary centrifuges in the laboratory could handle about 1 liter of culture at a time, with an hour of centrifuging being required to separate the cells adequately. This made it impractical to deal with more than a few liters of culture at one time.

An alternative device for the recovery of bacteria from large volumes of culture had become available by modification of a machine originally designed as a cream separator. The machine consisted of a balanced metal cylinder about one and three-quarters inches in diameter and ten inches long that was mounted in a vertical position, connected to a turbine driven by compressed air so that the cylinder could be spun at speeds of 30,000 rpm or higher. When whole milk was allowed to flow in at the bottom of the rotating cylinder, the cream and milk would be separated into layers in the centrifugal field as they rose in the cylinder and could be collected from separate outflow spouts at the top. The operation was even simpler as applied to a bacterial culture, since the particulate bacteria would stick to the wall of the cylinder and the depleted culture fluid would flow out at the top. It was feasible to pass large volumes of culture through such a machine in a reasonable period of time and leave all of the bacteria packed inside the cylinder as a paste with about the consistency of a yeast cake.

Obviously this cream separator–centrifuge was just the thing for dealing with mass cultures of bacteria, but it had one serious flaw when it came to applying it to pneumococci. In the course of its operation at high speed, it emitted an invisible aerosol laden with bacteria, which dispersed into the immediate environment. While this may have been tolerable in dealing with various nonpathogenic bacteria, it was totally unacceptable when one was centrifuging living, virulent type III pneumococci. Sometime before the fall of 1940—I have no record of the exact date—Colin MacLeod set about to find a way to overcome this defect so that the machine could be used with pneumococcal cultures. With the assistance of a mechanically talented technician he designed an airtight, sealed housing for the cream-separator-type centrifuge. This housing, constructed in the Institute's machine shop, was a metal structure about 3 feet in diameter with access to the centrifuge through a gasket-equipped door which was fixed tightly in place by a series of bolts like those used on automobile tires. In fact, an ordinary tire wrench was used to remove and tighten them. The turbine of the centrifuge was driven by high-pressure steam rather than compressed air, and after centrifugation of the pneumococci a short blast of the same steam could be used to sterilize the chamber and the external surface of the cylinder. The effectiveness of this protective housing was thoroughly tested by running nonpathogenic bacteria through the machine and placing exposed culture plates in various locations throughout the room. There was no evidence of leakage into the environment. It was a rather cumbersome gadget but served the purpose well.

The unsuccessful fluoride-treated extract of October 29, 1940, is the first record I have of this centrifuge—referred to as "the Sharples," from the name of the manufacturer of the cream separator unit—being used with the pneumococcus. Over the next few years, thousands of liters of pneumococcal culture were passed through the machine with only a few mechanical difficulties and no mishaps resulting from dissemination of live pneumococci. The increased yields of starting material had a major impact on progress of the work; one could now try a variety of fractionation and purification procedures without being limited by the amount of crude active extract.

In the course of their other experiments, Avery and MacLeod had another look at the several components of the transforming system, confirming the earlier findings and trying to establish a reliable and consistent means of quantifying the activity of the fractions of their transforming extracts. The type II R strain, R36, had retained its favorable characteristics, although they discovered that if it were not watched carefully it would throw off mutants during subculture that were totally resistant to transformation. They learned to recognize these mutants by the appearance of the colonies they formed when grown on blood agar plates, so that if the cultures were monitored carefully they could avoid having the resistant mutants dominate the population of R pneumococci. The transformable variants were redesignated R36A. Another component of the system, the complex beef heart infusion broth that they used as culture medium, was unpredictable in its ability to support luxuriant growth of pneumococci and to provide the right conditions for transformation. To achieve more uniformity in this regard, they found it useful to apply a procedure that MacLeod had introduced to remove sulfonamide inhibitors from pneumococcal broth: adsorption of the broth with activated charcoal, yielding a pale yellow medium that was much more reliable in the transforming system. When it came to the serum factor that had to be present for transformation to occur, they continued to find that the specimens of chest fluid that they obtained for this purpose varied in their effectiveness but that heating at 60°C for 30 minutes improved even the best of them.

Attention to all of these points—the state of the R36 culture, charcoal adsorption of the broth, and careful selection and heating of the chest fluid—helped to provide a better system but by no means eliminated the annoying variability encountered in the transforming tests. The notes are sprinkled with the description of experiments that failed because "the system was off." Sometimes this was due to a slip-up in handling one of the known components, but more often the responsible variable was never identified.

It may be useful to have some picture of how the transformation tests were carried out at this time. Generally, one part of heated chest fluid was added to three parts of charcoal-adsorbed broth, and the mixture was placed in small test tubes (called Wassermann tubes because they were originally introduced by him for his serological test for syphilis) in 2-cc amounts. Varying amounts of transforming extract were added to a series of such tubes. All were then inoculated with one drop (0.05 cc) of a 10^{-4} dilution of a culture of R36A—roughly about 2500 colony-forming units. After overnight incubation at 37°C, a pretty good indication of the results of the test could be gotten by simply inspecting the tubes. Because of the anti-R antibody present in the chest fluid, the R organisms would adhere to one another during growth and settle to the bottom of the tube in clumps, leaving a clear supernatant fluid. If transformation occurred, the newly formed encapsulated type III cells would be unaffected by the anti-R antibody (because of the presence of the capsule) and would grow diffusely throughout the culture. Thus, a quick look at the tubes was usually enough to determine which ones showed diffuse growth indicating that the change had taken place. In practice, the contents of each of the tubes were subcultured by plating on blood agar media in order to confirm that type III organisms were actually present. When things went well, it was a highly satisfying and conclusive test.

Nothing much can be gained by a chronological recital of attempts that Avery and MacLeod made to analyze and fractionate transforming extracts during that year. It will be more appropriate to highlight those efforts that represented significant advances in the project. It is not always easy to detect the reasoning behind some of the approaches they used, since the rationale of their experiments was not often discussed in their protocols. I have made some deductions of my own on these points, based on the internal evidence and what I was told later by Avery of the manner in which the problem had developed.

They usually started their fractionation attempts with type III extracts rendered free of detectable protein by repeated application of the Sevag procedure. Having removed one constituent known not to affect transforming activity, they could then concentrate on other components of the mixture. One of the most promising, and frustrating, of the fractionation methods that they tried involved the use of calcium salts. When a solution of calcium chloride was added to a transforming extract, sometimes a precipitate would form, sometimes not. If a precipitate did form, it would often contain most of the transforming activity of the whole extract. Furthermore, even if no precipitate formed with calcium alone, the addition of alcohol in amounts too small to have any effect in the absence of calcium would bring down a voluminous precipitate. This would also carry most of the transforming activity. They also found that these precipitates contained most of what they called the "nucleic acid" of the extract, based on a qualitative chemical test called the Bial reaction. The Bial test was quite specific for ribonucleic acid—or RNA—which was often referred to at that time as the yeast-type nucleic acid.

Looking over these experiments today with the benefit of hindsight, it is apparent that they were coming close to a means of getting purified preparations of the transforming substance. What obscured the results was the presence in their extracts of large amounts of RNA and type III polysaccharide, both of which were thrown out of solution under certain conditions by the calcium-alcohol treatment. The presence of RNA in the active calcium precipitates did lead them to retest the effect of the enzyme, ribonuclease. They had tried this earlier, but it was now available in essentially pure and highly active form as a result of work at the Princeton laboratories of the Rockefeller Institute by Moses Kunitz, who had recently crystallized the enzyme from extracts of beef pancreas. Treatment of a sample of one of the transforming extracts with this enzyme broke down much of the RNA into molecules that were small enough to be readily separable from the other components. This separation was achieved by a process known as dialysis in which the enzyme-treated extract was placed in a cellophane bag and soaked in a large volume of salt solution. The minute pores in the cellophane allowed the small molecules to pass out of the bag while retaining the large ones. By means of the Bial test it could be estimated that 75 percent of the RNA had been removed by this process, and yet all of the transforming activity had remained with the nondialyzable material inside the bag.

In addition to making it unlikely that RNA had anything to do with the transforming principle, these experiments carried out in January 1941 introduced an important new element into their analysis of the extracts. In the course of applying the Bial test to the various fractions obtained, they tried an additional test that they had not used previously: a test for deoxyribose that depended on a reaction with the organic chemical, diphenylamine. The diphenylamine test was not particularly new but had apparently not come to their attention previously. What they now found was that the extract, both before and after ribonuclease treatment and dialysis, gave a definitely positive reaction characterized by the development of a china blue color that was said to be typical of deoxyribose and thus also of DNA. In his comment on these tests, carried out on January 28, MacLeod wrote: "Thus it would appear as though these transforming extracts may contain a little desoxyribonucleic acid* in addition to the large amount of ribosenucleic acid present."

This first indication that the pneumococcus contained DNA came as something of a surprise. Knowledge of the occurrence and distribution of the nucleic acids in nature had not yet reached the point where one could assume that all living cells contained both RNA and DNA. Indeed, the notion had only recently been discarded that there were two general classes of nucleic acid: plant nucleic acid, as typified by yeast preparations, and animal nucleic acids, as typified by thymus and fish sperm preparations. A better picture had emerged with respect to the situation in animal cells, assigning DNA to the nucleus and RNA to the cytoplasm, but information about bacterial nucleic acids was still rudimentary. While MacLeod's further comments indicated some doubt about the specificity of the diphenylamine test for DNA, they gradually applied the test more frequently as these doubts were dispelled.

Meanwhile things were not going all that well in the preparation of extracts for fractionation. On February 13, for example, Ex1/41 was prepared by the standard method from 35 liters of type III culture, and after testing its activity

*Some readers may find it confusing to encounter two different spellings for some of the chemicals referred to in this book, especially those beginning with "desoxy" or "deoxy." "Desoxy" was the standard spelling for compounds of this class through the first half of this century, and it was changed to "deoxy" by an international nomenclature committee in the late 1950s. I have used the old spelling only when quoting old documents and have adhered to the revised form elsewhere, even though I do not believe that it was a necessary or inspired change. In either form the term merely implies the removal of an oxygen atom from the molecule; and deoxycholate is a modification of the bile salt, cholate, with one of its oxygens deleted. Deoxyribose is related to ribose through elimination of the oxygen attached to its fifth carbon atom.

MacLeod characterized it as "very weak." It was very weak, indeed, transformation being obtained only with 0.5 cc of the undiluted extract. In the end, these difficulties led to the radical step of abandoning the Alloway procedure altogether. On March 11, Avery tried his hand at an alternative approach with a small batch of type III pneumococci from 2 liters of culture. I think that it is likely that MacLeod was away from the laboratory at the time, since the experiment is one of the very few recorded entirely in Avery's hand. What he did was to heat kill the pneumococcal cells immediately after they had been recovered by centrifugation, the idea being to inactivate the T.P.-destroying enzyme before extraction began. The difficulty posed by this maneuver was that the heat-killed cells were no longer susceptible to lysis by deoxycholate, thus interfering with extraction of the cellular contents. Avery tried to circumvent this by shaking the suspension of heat-killed cells with a somewhat higher concentration of deoxycholate than needed for lysis, relying on the detergent properties of the bile salt to leach material from the cells. His tests for transforming activity showed that active material could be obtained in this way.

I suspect that there was some difference of opinion between Avery and MacLeod about the promise of this approach. MacLeod had inserted two pages in the notebook describing experiments carried out in 1935 in which he had tried a similar procedure with rather indifferent results. This clue, together with the evidence that Avery had taken the unusual step of initiating the new trial on his own, suggests that MacLeod was not very optimistic about solving their extraction problems this way. They compromised by undertaking an experiment on March 18 in which the pneumococci from 40 liters of culture were divided into two equal parts; Avery immediately heat-killed his share at 65°C for 30 minutes to repeat his pilot experiment and MacLeod processed the other portion by the usual procedure. They kept separate notes on the preparation of their extracts but then joined in testing them for activity.

In reviewing the protocols, I cannot see that Avery's extract of heat-killed cells was substantially more active than MacLeod's preparation from lysed organisms, although the titrations were not yet being carried out in a manner that would allow precise quantitative comparisons. On the other hand, Avery's extract obviously contained less total material and showed strikingly less reactivity with pneumococcal antisera. The implication was that one could obtain at least as much active transforming principle from the heat-killed cells while bringing along with it a smaller amount of the inert materials that they would have to get rid of in the process of purification. On March 25 they took the obvious next step of preparing an entire lot with heat-killed cells from 45 liters of culture. This was a success, and the Alloway procedure was not used again in the continuing drive to uncover the nature of the transforming substance.

Even though the results of the comparison of the two methods of extraction had not been dramatic, the laboratory notes of this period began to reflect a remarkable change in outlook of the research. Not only was the production of satisfactory extracts much more consistent, but it gradually became apparent that they were dealing with considerably more potent material. Rather than obtaining extracts that were considered potent if 0.1 cc was able to effect transformation, as defined in their earlier work, they began to encounter material that was active when 0.1 cc of a hundredfold dilution was tested. Occasionally transformation occurred even when 1:1000 dilutions were used. It was apparent that much of the active material must have been destroyed in the process of deoxycholate lysis and that they now had in hand a procedure that was providing considerably more favorable starting material for purification.

This new development was still in its preliminary stages when they had to prepare the annual spring scientific report for the Board of Scientific Directors, and consequently this change in approach to the extraction of the transforming principle was not described. In fact, their report in general was somewhat reserved, as if they were holding back a bit. They even gave their section of the report a rather cryptic title: "Studies on Capsular Synthesis by Pneumococci." The omissions in the report can only be explained by assuming that they considered the work incomplete and in need of confirmation before it could be described in writing, even in an internal document such as this. Thus, they say nothing about finding DNA in the extracts, nor do they mention the calcium-alcohol fractionation results. The failure to mention DNA becomes more pointed in view of the following comment which they made about the ribonuclease experiment during which the diphenylamine test was first used: "Extracts prepared in this manner [essentially the Alloway method] contain considerable amounts of nucleic acid. The latter substance may be almost completely removed by digestion with crystalline ribonuclease without affecting the transforming potency."

Much of the report deals with a line of investigation, carried on concurrently with the fractionation attempts, that involved further study of the various crude enzyme preparations that were able to destroy the transforming substance. The idea, of course, was to seek for clues as to the nature of the active substance by determining what components these enzyme preparations had in common; in other words, what kind of an enzyme was causing the destruction. They had turned up the information, more misleading than helpful in the long run, that each of the effective enzyme mixtures contained an esterase which they measured by its action on a low-molecular-weight ester (really in this case a simple fat, glycerol fully esterified with butyric acid). Their interest in this component was heightened by the fact that fluoride, a known esterase inhibitor, exerted some inhibitory effect on the action of the enzymes on the transforming substance. The following comment was included in the report: "From the evidence obtained by the use of various enzyme preparations it is not held that the specific transforming principle is necessarily of the nature of an ester, inasmuch as other enzymes may have been present. However, the evidence obtained by the use of sodium fluoride as an enzyme inhibitor when taken in conjunction with other studies, suggests that the transforming principle may be an esterified compound." It was never clear to me what kind of esters they had in mind, and there is nothing to indicate that they were considering the kind of phosphodiesters that hold together the backbone of nucleic acids.

Sometime earlier that year MacLeod must have been approached by New York University with an offer for him to assume the chairmanship of the department of microbiology at its college of medicine. I have no record of the date or the details of negotiations, but it must have disrupted progress in the laboratory work. The matter was settled by March 1941, since Tillett knew of Colin's impending move when the Fellowship Board of the National Research Council suggested that I go to Rockefeller to work with him. The prospect of this move was inevitably a blow to Avery and, as I found later, was not something that MacLeod anticipated with unalloyed enthusiasm.

Colin talked about this matter at some length in a late night conversation when we were sharing living quarters during a meeting at Yale University in 1959. He had had no desire to leave Rockefeller and, on the contrary, was hopeful of finding a permanent place there. However, when he went to Dr. Rivers, the director of the hospital, with the news of the New York University offer, rather than generating some counteroffer from Rockefeller, he was strongly urged to take the job. The implication of this kind of advice from the director was that the prospects were dim at Rockefeller and that he should grab a good opportunity when he had it. My own reconstruction of the situation was that Rivers had already selected the individual who would head a continuing pneumonia service at the hospital after Avery's retirement two years hence. This was Frank Horsfall, always a Rivers favorite, who had left the Avery laboratory in 1937 to join the Rockefeller Foundation laboratories that were maintained on the Institute campus where he was engaged in virus research. This fit in with Rivers's interests and his idea that studies on pneumonia in the future should focus on viral pneumonia. In any event, Horsfall was made a full member of the Rockefeller Institute as of July 1941 and rejoined the Avery laboratory just as MacLeod was leaving.

Thus, in a sense, Colin lost out to his old friend and colleague. He was not the kind of person, however, to be immobilized by disappointment, and he continued vigorous activity in the laboratory while preparing himself for the move to N.Y.U. and a rather different career. The research followed the path that had already been started, with some more experiments on the inactivating enzymes and their inhibition and a number of attempts to sharpen up the results of fractionation of the extracts with calcium and alcohol. A dozen additional large batches of type III pneumococcal cells were extracted during May and June to provide material for fractionation, although the final four were precipitated by alcohol and stored over the summer for future work. By and large, the results were so variable as to be encouraging one day and discouraging the next. The transforming activity might be largely concentrated in a single fraction in one experiment and then spread throughout a series of fractions when they repeated it. A given procedure would give promise of separating T.P. from other components, such as RNA and SSSIII, but behave quite differently when they tried it again. Looking back with the knowledge of subsequent events, I would guess that their difficulties came from the complexity of the mixture with its variable concentration of the several components, since the sharpness of precipitation with both alcohol and calcium depends on the relative concentration of these components.

It is not always easy to follow the details of the manipulations used in the fractionation attempts, but one significant new ingredient was introduced during this period. In an experiment carried out on May 12, MacLeod commented as follows: "It has been noticed previously that when alcohol is added to an extract, the precipitates formed are of 2 kinds—*a.* coming out at about ½–¾ volume of alcohol is stringy, veil-like; *b.* flocculent at higher concentrations of alcohol. In this expt., therefore, the attempt is made to obtain these 2 gross fractions separately." On this occasion most of the transforming activity and most of the type III polysaccharide (or SSS) appeared in the first type of precipitate and most of the RNA in the flocculent fraction. However, the results on further trials of this kind were inconsistent, and they obtained some suggestion of separation of T.P. and SSS in this process.

The notes make it clear that they were equating the stringy or fibrous alcohol precipitates with SSS and attributed the viscosity of solutions of the extracts to this component. Scattered throughout the descriptions of the procedures are comments such as: "viscous-like SSS," "precipitate stringy, fibrous—looks like SSS," and "Ppt. is flocculent with very few strings of SSS." They had a good reason for this assumption, since they had earlier obtained a quite pure preparation of SSS with these properties as a by-product of their fractionation efforts. Nevertheless, this notion tended to obscure the possibility that there might be another component of the extract with similar properties.

Just as he was departing MacLeod wrote on July 7, as a kind of distillate of the year's efforts on T.P., a detailed summary of the methods they had adopted for the preparation of extracts, and he included a flow sheet describing procedures for fractionation with calcium and alcohol. Since the latter was based on a series of rather inconsistent experiments, he had to hedge a little in defining what one could expect from the individual steps. Then at the end of his outline of the method he added two final sentences: "In this process of purification there has been a great loss of activity. It may be necessary to increase the concentration of SSS to improve the results." This latter comment is rather cryptic, since there is little in the notes to suggest that they had considered the type III polysaccharide of importance in the transforming reaction. As a matter of fact, it had long been assumed in the laboratory that type III SSS had nothing to do with it. I later learned what these second thoughts about the role of SSS in transformation were all about.

It had recently been discovered that an enzyme capable of synthesizing glycogen (a polysaccharide composed, like starch, of only glucose molecules) would not work unless some glycogen molecules were present in the reaction system. In other words, the enzyme had to be primed by some of the end product that it could build on during the synthesis. As an explanation for the confusing results and loss of activity encountered in their fractionation experiments, Avery and MacLeod had conceived the notion that something of the sort might be involved in transformation. Without weakening their conviction that the transforming principle was distinct from the type III polysaccharide, it seemed possible that the enzymes induced by T.P. in the process of transformation would not be able to initiate the synthesis of the polysaccharide and form a capsule unless there was an adequate supply of SSS in the system to serve as a primer. They were not confident that their tests of the role of the polysaccharide carried out years earlier had eliminated this possibility. The need to resolve this question provided me with my first project on transformation after I joined the laboratory that fall.

In essence, then, this was the state of the research on transformation when I arrived on the scene in September 1941.

VII

MY INITIATION

LABOR DAY fell on September 1st in 1941, forcing me to delay my long anticipated appearance at the Hospital of the Rockefeller Institute until Tuesday the 2nd. When I presented myself at the front desk, it was pretty obvious that the receptionist was not expecting me. After her call to the laboratory to confirm that I was a new postdoctoral fellow, it was Frank Horsfall himself who came down to greet me. He also seemed to be no more than vaguely aware that I was scheduled to join the laboratory, but he received me cordially and was most helpful in getting me settled and introduced to the other members of the group. Although I knew that Avery would not yet be back from Maine, I had seen no reason to wait for his arrival to begin the fellowship. As it turned out, I had an opportunity before his return to learn about the general operations of the laboratory and to become familiar with some of the Institute's other outstanding facilities: the superb library and the fabulous lunch room. It was in the latter that I was able to meet within a matter of weeks essentially all of the members of the Institute's scientific staff.

Nothing about the Avery laboratory would impress those who have depended on the movies for their idealized picture of a laboratory. It had no elaborate or mysterious equipment,

The Rockefeller Hospital. The Avery laboratories were on the sixth floor at the left (the east end of the building). *(From the Rockefeller University Archives.)*

and of course no automated apparatus, electronic gadgets, or computers. Even its structural features seemed anomalous, since the area had originally been designed as a hospital ward and was modified very little for its revised function. The original planners of the hospital had overestimated the amount of bed space that would be required in relation to research laboratory space. The number of patients need not be too large for intensive clinical research, and indeed if too many are involved the investigators may be so overwhelmed by purely clinical responsibilities that the research is shortchanged. In order to correct the imbalance, the sixth floor of the hospital, where the Avery laboratory was located, had been converted in the early years to laboratory space without any significant expenditure of funds. None of the partitions were changed, and the wards, private rooms, kitchens, etc., were simply adapted to their new function by moving in a different type of furnishings. The basic unit for each investigator was the microscope desk, a modification of the conventional knee-hole desk that had a microscope cabinet on one side. The standard accessories for bacteriological work at each desk were a Bunsen burner, a collection of bacteriological loops and needles made of platinum wire attached to long rods, and receptacles for the disposal of contaminated glassware.

The large room to which I was assigned had four of these desks. Mine was placed in front of the marble mantel of a fireplace, part of the embellishment of the original ward but of course no longer functional. Aside from an incubator, a sink, a fume hood, and a large square cabinet with a flat top surface in the middle of the room, there was not much else in evidence. The collection of centrifuges were all segregated in a small room in another area where their noise could be contained. Avery's private laboratory, in a small room adjoining the main lab, had apparently originally been a ward kitchen and was reached through a swinging door equipped with an oval window. His desk stood against a bank of north-facing windows but the view of the campus was usually obscured by a dark shade that was drawn to provide a better background for the examination of various test tube reactions with the aid of a green-shaded lamp that hung over the desk, suspended from the ceiling. An electric refrigerator of the household variety was the only item of equipment. On the other side of his laboratory, through another swinging door, was a small room modestly equipped for chemical work. Here I ultimately carried out the experiments for preparation of an enzyme that would break down DNA.

The stock-in-trade of the laboratory was sterile glassware, and the collection was quite large and diverse. It included flasks of different sizes, ranging from cumbersome ones of 4-liter capactiy that were used for growing the mass cultures of pneumococci, down to tiny flasks for analytical purposes; various kinds of test tubes, dominated by a large supply of Wassermann tubes for the transformation experiments; and graduated pipettes of 1-, 2-, 5-, and 10-cc capacity, each of which had been individually wrapped in brown paper before sterilization. The fractionation of the extracts also required a number of beakers, bottles, and other items for the "kitchen chemistry"—as Avery called it. Then there were a number of heavy glass centrifuge tubes and bottles, specially designed to withstand breakage under the stress of centrifuging. All in all, the place was set up in much the same way bacteriological labs had been for a long time.

By the time Avery returned in the middle of September, I was pretty well acclimated. Having read much more about the pneumococcus and the earlier work of the laboratory, I felt that I was prepared for his indoctrinating process. There were two other young investigators in the laboratory, Dick Mirick and Ed Curnen, who had been working with MacLeod for two years on certain aspects of the sulfonamide problem and were continuing their projects. They had given me some idea of what I was to expect from Avery as my introduction to the family and had warned me that I was not going to be assigned a problem. The decision on my project would emerge

O. T. Avery in 1941 at about the time the author joined his laboratory.

from talks with Avery and my reading and would be pretty much up to me.

The heart of the Avery process of orientation was a series of beautifully planned discourses that dealt with the major lines of his pneumococcal research. These discourses had come to be known by his young associates as Fess's "Red Seal Records," a term that reflected their high quality and also the fact that they tended to be repeated in much the same form on each occasion that he delivered them. It is not that they were memorized and repeated word for word, but certain phrases and descriptive terms had caught his fancy so that he used them consistently in the same way in telling a given story. He had preselected the most effective language and then stuck to it. There is no doubt that the Red Seal Records were effective. They were used not only for new arrivals in the laboratory but also for visitors and for young members of other research groups in the hospital. Many of the latter have vivid memories of these discourses, usually given in the evening or on a weekend when Fess was likely to be on hand and pleased to find a willing listener.

None of Avery's former colleagues have been able to explain his extraordinary success in training young scientists. Even though his research group was never very large, a remarkable number of those who passed through his laboratory emerged as leaders in medical microbiology. Certainly more was involved than the Red Seal Records, his philosophy of scientific research, and notions about how a laboratory should be operated. Some of his originality and creativity must have rubbed off on us as we learned from his experimental genius. The objective evidence for his success in turning out skilled investigators is clear. At least a dozen of his former "students" were subsequently elected to membership in the National Academy of Sciences, and this was at a time when the Academy had well under a thousand members in all fields of science, only a small fraction of whom were in the medical sciences. When I was elected to membership in 1963, the Section on Pathology and

Microbiology (which included most of those in medical science) had only forty-six members, and ten of them were former associates of Avery. His influence was also reflected in the composition of the staff of the Rockefeller Institute. In 1953, just as it was entering a period of rapid growth with the addition of a program in graduate education, there were only eighteen on the entire Institute staff with the top rank of Member, and four of these had come from Avery.

It was from one of Fess's discourses, shortly after his return from Maine, that I first heard the full story of the transformation phenomenon and learned what he and MacLeod had been up to for the past year. It captured my imagination from the beginning. I was soon certain that I wanted to work on the problem, but I wasn't sure how matters stood and was uncertain how to broach the subject. MacLeod had paid some visits to the laboratory beginning in the first week after Avery's return in order to join him in another crack at their fractionation procedure, using the four extracts that they had stored under alcohol over the summer. On one of these occasions, I can recall the three of us standing around that high cabinet in the center of the lab discussing possible research problems for my fellowship year. MacLeod enthusiastically suggested that I pursue further the work I had done with Tillett on the prevention of benzene leukopenia by sulfapyridine. Nothing could have been further from my thoughts.

I now realize that I was not sufficiently sensitive to MacLeod's feelings about the transformation problem throughout the early years of our association. I did not know then that he had left Rockefeller most reluctantly, nor was I yet aware of how large a stake he had in the development of the research over several years. It is even possible that he entertained some notion of taking the problem with him to N.Y.U. There was every reason to suppose on the basis of past performance that Fess would not pursue it on his own. MacLeod could not have looked upon my untimely arrival in the lab with other than mixed feelings. Yet he was always generous and helpful to me, and our friendship never appeared to be poisoned with any taint of resentment. Many years later, after his death, I received the first hint that he may have been less than pleased with my participation in the research.

The efforts of MacLeod and Avery that September to sharpen up the results of their fractionation procedure met with the same kind of variability that they had encountered earlier. MacLeod was clearly only present for short periods during these experiments, but he wrote the notes for the fractionation attempts. At the end of his summary, which represented a flow sheet for the procedure, he added the following note: "most of the activity should be in Ppt. #4." Unfortunately, from the transformation tests carried out by Avery, it was evident that "Ppt. #4" had very little activity and that most of it was present in another fraction that was designated SSS. None of the fractions gave a diphenylamine test, presumably because they were too dilute. It was obvious that the procedure was giving unpredictable results and still left them with some apparent ambiguity concerning the part played by SSS.

Avery continued to carry out experiments with the extract prepared that September, testing the various fractions and trying to sort out the confusing results. Then one day late in the month, while I was still undergoing my informal indoctrination, he suggested that since I seemed interested in the transformation problem I might like to join him in setting up one of the tests and observe the phenomenon at first hand. Discussion of this experiment and its implications led to my participation in other tests, and without any explicit agreement about my role I quickly slipped into the position of his full-time collaborator in the research. From the second week of October onward all of the laboratory notes are in my hand with only occasional entries by Avery. Thus, the usual process of selecting a research problem had been accelerated in my case, stimulated by my obvious interest in the search for the identity of the transforming principle and Avery's pressing desire to get on with it.

Avery was an active participant in the laboratory work, even though I soon found out for myself that he was not inclined to initiate a new experiment. After we talked over what steps to take next and agreed on a course of action, it was up to me to set up the protocol and assemble the necessary materials. However, he would then review the protocol and join me with enthusiasm as we sat side by side to set up the experiment. Often this involved a large number of test tubes to accommodate the several variables being tested, in addition to a number of controls. It was in this manner that I was introduced to Avery's extraordinarily rigorous bacteriological technique. He was fond of telling how he and Ben White, his associate at the Hoagland Laboratory in his pre-Rockefeller days, had agreed that they would treat all bacterial cultures as though they contained the plague bacillus. They realized that it was a common failing to become sloppy in handling nonpathogenic organisms which in turn led to some relaxation of acceptable techniques when dealing with more infectious agents. Avery adhered for his entire career to this early resolution to use maximum care in handling all bacteria.

In practice, his technique involved a series of rituals for such procedures as unwrapping sterile pipettes, flaming the bacteriological loop, or manipulating the cotton plugs of sterile tubes or flasks. An experiment was not begun until the required tubes, pipettes, reagents, and racks were systematically arranged on the desk for ready accessibility, and the Bunsen burner properly positioned. He would then draw the chair close so that the right hand, holding the pipette, could be stabilized by placing the right elbow firmly on the desk. The pipette, containing such material as the sterile medium or bacterial culture, would then be held nearly stationary, with the tip one or two inches from the flame of the Bunsen burner. The left hand would be used to move tubes and flasks to the scene of the action, bringing them first to the fourth and fifth fingers of the right hand for removal of the cotton plug, then to the burner for flaming of the opening, to the pipette tip for delivery of the sample, back to the burner for reflaming, and then to the right hand to retrieve the plug. All this was done with almost no movement of the pipette. A break in technique, such as touching the pipette to the outside of a tube or brushing a hand against an exposed cotton plug, resulted in immediate discarding of the possibly contaminated material. As a result of these precautions, chance contamination during the setting up of an experiment was virtually eliminated.

Although I soon began experiments directed toward the major goal of the research on transformation, an important prerequisite was to learn the procedures for preparing extracts from the mass cultures of type III pneumococci. On October 21 Colin MacLeod came to the laboratory to guide my hand in my first attempt to cope with a batch of organisms from the Sharples centrifuge. As was the custom, the cultures were handled by Fred Kimmer, an elderly technician who had been at the Institute for many years and at one time had worked with Noguchi. He inoculated 51 liters of culture medium (seventeen 4-liter flasks containing 3 liters of medium each) with a heavy inoculum of type III pneumococci early in the morning so that they would be ready for harvesting by mid-morning. The culture was chilled as it ran into the Sharples by passing it through copper tubing immersed in an ice-water bath in order to terminate growth and protect it from possible overheating during the final steam sterilization of the outer centrifuge chamber. After completion of the centrifuge run, the cylinder containing the packed organisms was removed, wrapped in a Lysol-soaked towel, and delivered to the laboratory. Then the fun began.

The problem was to get the bacterial cake out of the cylinder and into suspension in salt solution without spreading viable organisms all over the place. Teddy Nadeje, the same technician who had aided MacLeod in devising the housing for the centrifuge, had fashioned an instrument for scooping the organisms out of the cylinder. It was a half-round, thin metal plate machined to fit the inner surface of the cylinder and attached to a long metal rod. The bacterial cake was transferred to a beaker on this gadget where it was scraped off with the aid of a spatula. After several repetitions of this procedure and some washing with salt solution, the material was all in the beaker ready to be smoothly suspended in more salt solution and promptly heat killed. It was, however, a messy procedure. No matter how careful one tried to be in scooping the bacterial cake from the cylinder, there were bound to be little slips and sudden jerks. As a result, one would see small flecks of white material fly in one direction or another with the disconcerting awareness that they were composed of millions of viable pneumococci. Despite all precautions and the liberal use of germicides, one could not complete the task without the conviction that he had thoroughly contaminated himself and the immediate environment.

Following my lesson from MacLeod on how it was done, I processed a large batch of pneumococci on the average of once a week for the next few months. Each of these was extracted by shaking the suspension of heat-killed organisms in the presence of deoxycholate and, after removing the organisms by centrifugation, the extract was precipitated by alcohol. The total precipitate was then redissolved in salt solution for removal of protein by repeated application of the Sevag method. At this point, most of these early extracts that I prepared were freeze-dried, a procedure that effectively preserved the activity of the crude extracts so that they could be reconstituted as needed later for fractionation experiments. Material from several hundred liters of culture had been stockpiled in this way by the end of January 1942.

As I was handling these "Sharples runs," as we called them, I gradually became aware that Fess was never present in the laboratory while the organisms were being removed from the Sharples cylinder and prepared for heat killing. If he happened to be there when the cylinder was brought up from the centrifuge room, he would quickly disappear. While I could give no credence to Fred Kimmer's view that this behavior was motivated by fear of possible infection, it took me some time to realize what the answer really was. It was simply that he could not bear to witness a procedure that deviated so far from his standards of correct bacteriological practice. He accepted its necessity for the research, but he could not be a party to it. As soon as the bacterial suspension had been killed by heating at 65°C for 30 minutes, he was ready to take part in all of the subsequent steps.

Concurrently with the business of making transforming extracts, I had begun analytical studies to reexamine the possible part played by the type III capsular polysaccharide in transformation. Was it really necessary to have a little SSS around as a template or primer in order for the transformed cell to start synthesizing the polysaccharide so that it could form a capsule? With the availability of the Dubos SIII enzyme this would seem to be a simple question to answer, and indeed the enzyme had been applied to the problem earlier, although I was not aware of this at the time. Both Rogers in 1933 and MacLeod in 1935 had tested the effect of the enzyme on their transforming extracts and found that it did not seem to affect transforming activity. The experience that MacLeod and Avery had with their fractionation experiments in the spring of 1941 raised some doubts in their minds whether these early tests had been sufficiently quantitative and complete enough to answer the question. The idea was to be certain that the SIII polysaccharide was completely eliminated by the enzyme and that even the split products of the enzyme's action were removed.

Dubos, who still had some of his original preparations stored in the freeze-dried form, supplied the enzyme for my attempts to deal with the problem. The activity of the enzyme could be demonstrated quite simply by adding some of it to a solution of the type III polysaccharide and then testing samples at intervals for their ability to give a precipitate with type III antiserum. A good enzyme preparation could destroy the precipitating reactivity of a dilute solution of the polysaccharide in less than an hour. Since our type III antisera would form a visible precipitate with solutions of polysaccharide at concentrations as low as 0.0002 milligram per cubic centimeter, the test had a good degree of sensitivity. As I began experiments on the effect of the enzyme on transforming extracts, I was treated to a thorough indoctrination on the vagaries of the transforming system. I encountered all of the difficulties of the past and learned the methods of checking the medium, the chest fluid, and the R strain as possible sources of the trouble. After one experiment in which there was no growth in any of the tubes, I inspected the test tubes for incomplete removal of the detergent used in washing the glassware. It was necessary to consider all possible sources of trouble and to be persistent until the problem was solved. A great deal of repetition of experiments was required.

At the end of November I finally completed a protocol experiment that was satisfactory from all points of view. This showed that if one treated a transforming extract with the enzyme until it no longer had any serologically detectable type III polysaccharide, and then thoroughly dialyzed away the split products of the enzymatic reaction, there was no loss whatever of transforming activity. Although I did a few more experiments of this kind to substantiate the finding, the results were quite conclusive in establishing that SSS did not have to be present for transformation to occur. It was time, therefore, to get on with an effort to isolate and identify the transforming substance.

One of the clear implications of the SIII enzyme results was that the type III polysaccharide was something that we should strive to get rid of in the process of purification of T.P., just as we did with protein. The enzyme was an obvious candidate for the job, but the supplies then available, while quite adequate for the analytical studies, were not up to tackling the large amounts of polysaccharide contained in whole extracts. This led us to consider other possible means of reducing the amount of polysaccharide present in the extracts, to a point where the residual remaining could be removed by the limited amounts of enzyme. One such possibility that suggested itself was to modify the conditions under which we grew the type III pneumococci for extraction. In the procedure we were then using, Fred Kimmer would add a generous shot of glucose to each flask to enhance growth during the last 2 hours of incubation. This certainly achieved its purpose, but at the same time the glucose, which the organisms used for the synthesis of polysaccharide as well as for metabolic energy, greatly increased the amount of capsular material produced. We decided to try using a smaller initial inoculum of type III pneumococci and incubating the cultures overnight without the addition of excess glucose.

When the next extract was prepared after the successful SIII enzyme experiment, on December 2, we adopted this modified procedure. As expected, the bulk of packed cells obtained after centrifugation was smaller than usual, and correspondingly the amount of material extracted seemed to be less, as indicated by the appearance of the alcohol precipitate. Compensating for the reduced yield, serological analysis showed that the extract had only a fraction—not more than 10–15 percent—of the type III polysaccharide found in an extract prepared two weeks earlier by the old method. The real surprise came when we tested its transforming activity. It was more potent than any previous extract. Transformation occurred in tubes to which we had added as little as 0.3 cc of a 1:10,000 dilution. Two repetitions of this procedure confirmed that we could obtain highly active material containing substantially less SSS in this way, and we abandoned the use of extra glucose from then on.

I had driven into the laboratory on Sunday, December 7, to read the agar plates from the transforming test of the promising extract prepared earlier that week and to set out some cultures for the next day. On the way home I heard news of the Japanese attack on Pearl Harbor on the car radio. The turbulent world situation, disturbing enough before this, was now even closer to home. Not immune to the surge of patriotic fervor, I found it difficult at first to concentrate on laboratory work. Dr. Rivers had had the foresight to make some preparations for war as far as the Rockefeller Hospital was concerned. A year or so earlier he had organized a U.S. Naval Reserve unit for medical research to be based at the hospital. If the unit were called to active duty, it was his intention to use the hospital for special studies of medical problems that arose among the navy personnel. He had recruited into the naval reserve most of the younger men on the hospital staff, including Horsfall, Mirick, and Curnen of the Avery laboratory. When I went to Rivers shortly after Pearl Harbor and asked to join the reserve unit, he declined my request. His reasoning was that I was married, with young children, and thus unlikely to be drafted. He suggested that I carry on just as I had been.

Without being wholly comfortable with his decision, I managed to throw myself back into laboratory work. My efforts during the succeeding weeks centered on trying a number of different ways of separating the type III polysaccharide and RNA from the active transforming principle in the hope of finding methods that might be applied to the purification process. Some of these had been tried before by MacLeod, and I didn't have any more luck than he had in hitting upon a selective method. It did occur to me, however, that in addition to limiting the amount of glucose available to the pneumococci during growth it might be worthwhile to wash the organisms before extracting them with deoxycholate. Since the polysaccharide is on the surface of the bacterial cells, washing the cells offered some promise of removing a large part of it and possibly even some of the RNA. Up to that time the routine procedure had been to add deoxycholate to the bacterial suspension immediately after heat killing and to proceed directly with extraction. In pilot experiments with small batches of pneumococci, I centrifuged the organisms after the heat-killing step and washed them with more salt solution before submitting them to deoxycholate extraction. Tests of the individual fractions revealed rather unexpectedly that there was readily detectable transforming activity both in the supernatant of the heat-killed cells and in the saline wash. There was much more in the deoxycholate extract, however, and any losses encountered were more than compensated for by the marked reduction in the amount of type III polysaccharide and RNA present in comparison with that in the supernatant of the heat-killed cells. Later experiments with large batches and more quantitative titration methods showed that more than 90 percent of the transforming activity could be recovered in the deoxycholate fraction even after repeated saline washing of the cells before extraction. We were on the way to a more rational set of extraction procedures.

During the early weeks of 1942 I also pursued further the possible use of the SIII enzyme in purification of the transforming substance. While I had to come ultimately to the preparation of my own enzyme for this purpose, I still had enough from Dubos at this point for a number of trials. An extract prepared on December 16 from cells grown without extra glucose had been stored under alcohol after deproteinization rather than dried from the frozen state. This material was redissolved and subjected to further purification attempts using both the SIII enzyme and ribonuclease. After repeating the Sevag process to remove the added enzyme protein and dialyzing the solution to eliminate split products, it was precipitated with alcohol. To our surprise, before one volume of alcohol had been added, a stringy mass of fibrous precipitate separated out. Clearly it was not only the type III polysaccharide in our extracts that could yield alcohol precipitates with these properties. Furthermore, after crude separation of this fibrous precipitate from the more flocculent precipitate produced by higher concentrations of alcohol, transformation tests showed that 99.9 percent of the activity was in the fibrous fraction.

At the time these observations were made in January 1942, there is no indication in the notes that we equated the fibrous alcohol precipitate with DNA. Curiously, we did not even record any diphenylamine tests for deoxyribose on the fractions. The emphasis of our tests was solely on protein, RNA, and SSSIII—the components of the extract that we were trying to get rid of and thus needed evidence for success in eliminating them. Nevertheless, this experiment marks the beginning of a period when our attention was focused with increasing sharpness on the possibility that the transforming principle might indeed be DNA. I have tried as best I can to reconstruct the sequence of events during this period in an attempt to sort out the various factors that influenced us to direct our primary attention to DNA.

VIII

INTIMATION OF SUCCESS

THE PROCESS leading to our serious consideration of DNA as the bearer of transforming activity was surely gradual. Nothing in my memory or in the laboratory notes suggests that there was a moment of sudden revelation, a single experiment that resulted in a flash of insight and reorientation of our thinking. On the contrary, the results of several different experiments and the injection of some new information from outside the laboratory were all involved in the crystallization of the concept. Almost certainly the notion that DNA might be the transforming principle had been entertained from time to time after MacLeod and Avery had detected the presence of DNA in their extracts a year earlier. It was an attractive possible identity for a substance capable of causing predictable and heritable changes in pneumococci in view of the known localization of the DNA of higher organisms in the chromosomes, the heart of the genetic apparatus. It had not been possible to design experiments that would test this notion directly, however, and none of the data obtained during the previous year threw much light on the problem. By January 1942 we were aware that the transforming principle was usually associated with material that gave a stringy, fibrous precipitate with alcohol—even when there was no type III polysaccharide present—and before long we recognized that most of the DNA in our extracts occurred in this same fibrous fraction. At about this same time we had an opportunity to learn more about the appearance and properties of pure DNA as isolated by an improved technique from mammalian tissues. This was new to us, since the sample of thymus nucleic acid that we had obtained as a standard for the diphenylamine test had been extracted with strong alkali, and it was a brownish powder without any of the characteristics that we later knew to be typical of native DNA. This sample had come from a collection of chemicals that had accumulated in the laboratory of P. A. Levene, a Rockefeller biochemist who had carried out much of the structural work on the nucleic acids.

Two floors above us, at the top of the hospital building, the biochemist, Alfred E. Mirsky, had been for some time carrying out basic studies on proteins. He was associated with the laboratory of Alfred E. Cohn whose clinical studies were concerned with heart disease, and as a result Mirsky's efforts had been directed toward studies of myosin, the principal protein constituent of muscle, including heart muscle. In the spring of 1941, his findings with myosin led him to take a look at another substance with similar properties that had been described a few years earlier and given the name "plasmosin," indicating its supposed origin from the cytoplasm of liver cells. In the course of purifying plasmosin, Mirsky found that it was quite different from myosin, being composed of a complex of nucleic acid and protein, with the nucleic acid component being DNA. In collaboration with the cytologist, A. W. Pollister of Columbia University, he showed that this nucleoprotein was in reality derived from the nucleus of tissue cells, and they ultimately gave it the name "chromosin."

Mirsky and Pollister refined the methods that had been used in extracting material like plasmosin into an elegant procedure for obtaining pure preparations of DNA nucleoprotein from any tissue. Simple and elegant, the method required little more than the use of salt solutions of different strengths. Organs like liver, spleen, or thymus were blended repeatedly with physiological salt solution—so called because it is compatible with blood and tissues—which dissolved away most of the soluble material of the cells and left behind an insoluble residue composed largely of the cell nuclei. When this nuclear residue was then placed in a much stronger salt solution that contained about seven times the concentration of sodium chloride used to make physiological salt solution, much of the material began to dissolve immediately to form highly viscous solutions of the nucleoprotein. The nucleoprotein could be precipitated from this solution in the form of fibrous strands simply by pouring it into several volumes of distilled water to reduce the salt concentration. Repetition of this process of solution in strong salt and reprecipitation by dilution led to purified products from which essentially all of the cellular constituents other than nucleoprotein had been removed.

The DNA and protein in these preparations are not linked together by firm chemical bonds but are attracted to each other by bearing opposite charges, the DNA being very acidic and the protein (called histone) very basic. The attraction between them is counteracted by high concentrations of sodium and chloride ions, which is the reason that strong salt solutions are able to dissolve the complex. To obtain pure DNA from this mixture, Mirsky had only to apply the same chloroform deproteinizing method of Sevag that we had been using on the transforming extracts. It took some doing to achieve this because the amount of protein was so great, but once it was all removed there remained beautifully clear solutions of DNA that no longer depended on high salt concentration for their solubility. The DNA could be precipitated from solution by alcohol to yield the same kind of fibrous mass that had become so familiar to us with our pneumococcal material, and after drying from pure alcohol and ether it appeared as a white bundle of tangled fibers that bore some resemblance to asbestos.

Sometime in the late winter or early spring of 1942 Mirsky gave us some preparations of his mammalian DNA. We were thus able to look at the characteristics and properties of this material in comparison with the fractions of our pneumococcal extracts. He also told us of his finding that if, during the process of precipitating his nucleoprotein by pouring the viscous solution into water, he stirred the mixture, the fibrous precipitate would wind around the stirring rod so that it could be simply lifted out as a single mass. It had been shown much earlier by a Swedish worker, Hammarsten, that the same thing would happen when pure DNA was precipitated by pouring solutions into alcohol. On March 30, I did a small experiment that I entitled: "The Nature of the Material in Transforming Extracts Giving 'Stringy' Precipitate in Alcohol." The notes of this experiment begin with the following rationale: "A method of preparing thymus-type nucleic acids depends upon the fact that when a solution of the material is poured into alcohol in a thin stream with constant stirring with a wooden rod, the nucleic acid collects as a stringy mass around the rod. There is apparently some thymus-type nucleic acid present in transforming extracts, and there is also material which gives a stringy precipitate with alcohol. The present experiment is carried out to determine the relationship of the 'stringy' material to the thymus nucleic acid in the extracts and also to the transforming activity."

I used a solution of one of the freeze-dried preparations of T.P. that had been stored away the previous fall. When the solution of this crude material was poured into alcohol, however, the results were not so tidy as I would have liked. While there was some fibrous precipitate that wrapped around the stirring rod, similar fibers of precipitate could be seen floating in the solution. The material was just a bit too crude to give a sharply defined separation. Nevertheless, more than half of the DNA, as indicated by the diphenylamine test, and more than half of the transforming activity were found in the fraction that collected on the rod. Most of the RNA remained with the material that had not adhered to the rod. The results were clearly in the right direction and strengthened our growing suspicions that we might be dealing with DNA as the transforming principle. The test also served as the model for the technique that we later used in modified form in the final purification of transforming DNA.

Much of our effort during the next month or two was devoted to an entirely different approach that served to focus our attention even more sharply on DNA. The ultracentrifuge had recently been developed into a powerful tool for the analysis of biological materials, based on the differential rate of sedimentation of macromolecules of different sizes. We were given an opportunity to apply this kind of analysis to our transforming extracts through the generous cooperation of Alexandre Rothen, one of the physical chemists on the Institute staff. Rothen's instrument, which had been designed and built at the Institute, was housed in a basement laboratory and took up most of the space in a medium-sized room. The centrifuge itself required only a small corner, and the dominant feature of the setup was the optical system used to visualize the rate of sedimentation of components of the sample during centrifugation. With this device any macromolecular substance present in adequate concentration could be detected as a moving boundary while the sample was spinning at very high speeds. The container in which the material was centrifuged was tiny, holding only about one-half cubic centimeter, but designed so that samples for chemical and biological analyses could be accurately removed after completion of the centrifuge run.

We began our ultracentrifuge experiments in mid-April, using one of the preparations of transforming substance that had been purified by removal of both protein and SIII polysaccharide. Very quickly we learned that the active substance must be an exceptionally large molecule and that it was not present in very high concentration, since it was sedimented more rapidly than the material that gave the fastest-moving visible boundary. Even at the relatively moderate speed of 30,000 rpm, only 1 hour of centrifugation was required to concentrate 99 percent of the activity in the lower third of the chamber. The only other known component of the extract that was similarly concentrated under these conditions was DNA. Here, then, was totally independent evidence to suggest that transforming activity and DNA were somehow associated.

The results of our several analytical centrifuge runs led Rothen to suggest that it might be useful to apply the machine to the purification of transforming extracts by using what he called the "concentration head." This would hold a number of plastic centrifuge tubes and allow much larger volumes to be spun at speeds comparable to that used with the analytical chamber. We had on hand many of the freeze-dried lots of transforming extracts that we had stored away the previous fall, so that it was a simple matter to redissolve and pool some of these to try out Rothen's suggestion. I was really not prepared for the results of this experimental approach. I was quite accustomed to spinning down bacteria and various types of precipitate, but it had never occurred to me that it was possible to spin dissolved substances out of solution. This was, in effect, what we encountered. After centrifuging our solution of deproteinized, but otherwise quite crude, transforming extracts at 30,000 rpm for a few hours, we found at the bottom of each of the centrifuge tubes a translucent, gelatinous pellet that remained behind after pouring off the supernatant fluid. The pellets could be readily redissolved in salt solution and shown to contain essentially all of the transforming activity and the DNA of the original solution. At the same time most of the RNA, the SIII polysaccharide, and other pneumococcal antigens remained in the supernatant fluid. While it seemed impractical to use the ultracentrifuge as a tool for the routine purification of T.P., the findings added an additional strong piece of evidence that we should focus our efforts on DNA.

The requirement for accurate estimates of the transforming activity in the various fractions obtained in the ultracentrifuge experiments made me decide to try to beef up the method of titration so that we could get more reliable quantitative data. My exposure to some of the work going on by this time in Horsfall's virus laboratory suggested that I might adapt the procedures used in measuring virus infectivity in mice to our system. Because of the biological variability of the response of even inbred mice to infection with an agent such as the influenza virus, it was necessary to inoculate several mice with each of a series of dilutions of the virus preparation. It was then possible to calculate with the aid of a statistical method the minimal amount of the virus that was able to infect 50 percent of the mice—a 50 percent endpoint for the titration. I felt that the tubes we used to carry out the transformation reaction, while clearly less complicated than the mouse, were showing the same kind of variability. Thus, by using four or more replicate tubes with each dilution of a solution of transforming substance, I could also calculate 50 percent endpoints. This proved to be effective, and as long as the system was working well I could get a good estimate of the relative content of T.P. in each of the fractions that were recovered from Rothen's centrifuge tubes. This became a routine method of assay, and it was applied to a variety of other fractionation experiments as well as to the measurement of the activity of the final purified preparations.

All this was going on amidst some further distractions occasioned by the war. In March, despite Dr. Rivers's assurances to the contrary, I received my draft notice for induction into the military service as a medical officer. When I brought this to his attention, he immediately reversed his earlier decision and initiated the process of having me added to his naval research unit, which had by that time been called to active duty. On May 6, 1942, in the middle of the ultracentrifuge experiments, I was commissioned as a lieutenant, j.g., in the U.S. Naval Reserve and assigned to active duty with the research unit. There was a slight hitch at the end when my orders came through assigning me to the Naval Hospital in Annapolis, but Rivers had enough influence to get these orders changed at the last minute.

For the next four years I worked on pneumococcal transformation in navy uniform. This anomalous position caused me a considerable amount of uneasiness and mental conflict. On the one hand, the research had reached such an exciting stage that it would have been extremely painful to give it up, but on the other there was a persistent, nagging feeling that I should be doing something more directly relevant to the war effort. Shortly after I was commissioned I went to Rivers with this problem, suggesting that it might be more appropriate if I were assigned to another project, such as the one that Horsfall had begun on virus pneumonia, a disease that was already a problem among military recruits. His reply was immediate and unequivocal: "No, you keep on working with Fess. That study is too important to drop. You don't have to worry about it." I still did worry about it from time to time, although the increasing tempo of the research and the excitement that was building up as all roads seemed to lead to DNA tended to keep my mind occupied with other matters. Later, after the first phase of our work on transformation was nearly completed, I participated to a minor degree in the studies of virus pneumonia and also took my regular turn "on call" as one of the medical officers in the hospital. Nevertheless, most of my time and effort during this whole period was devoted to one aspect or another of the problem of pneumococcal transformation. The navy regularly received summary reports of these studies, which must have mystified anyone who happened to read them, but no comment or questions came back to us. There is no indication that the navy ever recognized the role that it played in this early chapter of the biological revolution.

An unrelated outcome of my going on active duty in the navy was that for the first time in my life I was financially independent. One *could* support a family on the navy salary.

The time for preparing the annual spring report to the Board of Scientific Directors came just as we were immersed in the ultracentrifuge experiments. The results of these experiments were obviously not ready yet for reporting. Avery elected to avoid any discussion of our attempts at purifying the transforming substance, since it seemed likely that the situation was going to become much more definitive in the near future. He wrote an eight-page mini-treatise on pneumococcal virulence which included a description of three separate pieces of recent research on the subject: one carried out by MacLeod during the previous year, one by Horsfall in the period before he turned all of his efforts to virus pneumonia, and one by me as a sort of sideline of research during the first several months. Together they emphasized the complexity of the property of virulence and showed that while the presence of the capsule was essential for the expression of virulence it was not by itself sufficient. Other properties of the pneumococcus determine its ability to thrive in host tissues. One can under certain conditions obtain, for example, fully encapsulated type I organisms that are essentially nonvirulent. In addition, capsules of different serological types appear to afford varying degrees of protection to the organism. Thus, in contrast to types I, II, and III, type XIV pneumococci display little virulence for mice, even though they are commonly associated with pneumonia in man. MacLeod's and my work on virulence involved the application of the techniques of transformation and was thus not totally separate from our main line of endeavor. None of this was ever published except in the form of a brief abstract,[1] but it illustrates the continuing concern with the disease-producing aspects of the organism in the course of the pursuit of the nature of the transforming substance.

While we were mulling over the results of the ultracentrifuge experiments, it occurred to us that there was an additional way in which we could test our suspicion that DNA was involved in transformation. Since we knew of crude enzyme preparations from several different sources that would destroy transforming activity, why not examine the effect of these same enzyme preparations on authentic DNA, such as the material that Mirsky was making from mammalian tissues? Over the next few months, I set up a series of experiments repeating the studies of the action of the enzymes on T.P. and at the same time comparing their ability to degrade calf thymus DNA. The results were unequivocal; all of the preparations that destroyed transforming activity were also able to degrade DNA. Some additional pieces of information further strengthened the correlation. We knew that the effect of dog and human sera on the transforming substance could be eliminated by heating at 60°C but that this was not sufficient for rabbit sera, which required heating to 65°C. Tests of the DNA-splitting activity of sera followed precisely the same pattern, and rabbit sera could be inactivated only at 65°C or above while heating at 60°C was sufficient to eliminate the activity of dog and human sera.

The combination of these several lines of experimental evidence that consistently pointed to DNA had by the summer of 1942 pretty much convinced us that in all probability it was the transforming substance. We were not unaware that this idea would be greeted with skepticism and that we would need much more rigorous proof before we could consider publishing anything about it. We had already been told by more than one person, Alfred Mirsky being the first, that the transforming principle could not possibly be deoxyribonucleic acid because "nucleic acids are all alike." This point of view was widely held, having been generated for the most part by the work of P. A. Levene on the structure of the nucleic acids. For large molecules they were certainly deceptively simple in composition. The basic small molecular building block is called a nucleotide and consists of a nitrogen-containing organic "base," combined with a sugar (ribose in RNA and deoxyribose in DNA) and phosphate. Since there are only four different bases in a given nucleic acid, and thus only four different nucleotides, there appeared to be limited possibilities for diversity. Furthermore, Levene had advanced a "tetranucleotide theory," which proposed that the repeating structural unit was represented by the four different nucleotides hooked together in the same order, thus reducing drastically any chance for diversity among macromolecules of nucleic acid. There wasn't much solid chemical evidence for the tetranucleotide theory but it had nonetheless gained a fair degree of acceptance among biochemists.

Even if one did not accept this restricted view of the structure of DNA, the composition of the other major macromolecular constituent of chromosomes, protein, seemed much more favorable for the task of expressing genetic information. In the case of proteins, the small structural units are called amino acids, and there are twenty chemically different amino acids involved in the construction of a protein molecule. Since a moderate-sized protein contains hundreds of these amino acids linked together in a linear chain, the possibilities for diversity are almost infinite. It is no surprise, therefore, that in the minds of those who thought about the possible chemical nature of the gene, protein was the prime candidate. It was easy to find comments like the following from even the most eminent workers in the field: "If one assumes that the genes consist of known substances, there are only the proteins to be considered, because they are the only known substances which are specific for the individual."[2]

Not quite everyone jumped to this conclusion, however, and we were able to find some comfort in the writings of a few workers, even though they were unable to present any experimental evidence for diversity of the nucleic acids. In one of the few papers that directly addressed the problem of the chemical nature of the gene, Jack Schultz had considered the paucity of information on the "monotonous uniformity" of nucleic acids and wrote in 1941: "When it is considered that the highly polymerized thymonucleic acid has been studied in detail from a single source, and only recently have the ribose nucleic acids begun to be prepared in a comparably elegant manner, it is evident that the earlier conclusion can be accepted only as a first order approximation, and that much new data is necessary before we can exclude the possibility of specificities in the nucleic acids themselves."[3] Though hardly the kind of comment that could be cited as support for our growing belief that the transforming substance might be DNA, it was still something of a morale builder in the beginning stages of our flirtation with this idea.

Perhaps my favorite among the handful of quotes we were able to find expressing some optimism about the biological role of the nucleic acids I encountered in an old biochemistry textbook by R. A. Gortner.[4] I had come across this book while browsing in a bookstore in Chicago during my Stanford years and picked it up because it seemed to present the subject in a different manner from the text that we were using. Gortner had obviously been deeply impressed by a paper by the biochemist J. B. Leathes, entitled "Function and Design," which had been published in *Science* in 1926.[5] To the footnote marking the first citation of the paper, Gortner had added the comment: "Every biochemist or biologist interested in vital phenomena should read this paper." He quoted at length the calculations of Leathes that dramatized the vast diversity of the proteins; but later at the end of his discussion of nucleoproteins, after noting that nucleic acids form approximately 40 percent of the chromosomes, he returned to another statement of Leathes: if we consider that into these chromosomes "are packed from the beginning all that preordains, if not our fate and fortunes, at least our bodily characteristics down to the color of our eyelashes, it becomes a question whether the virtues of the nucleic acids may not rival those of amino acid chains in their importance." This was pure speculation, of course, but put forward with enough verve to have a special appeal for me.

As we were considering in the early summer of 1942 what steps would be necessary to verify the identification of the transforming substance as DNA, we became involved in a collaborative experiment with Mirsky that added some additional information on the subject but was probably more important for the ultimate impact that it had on the general acceptance of our work after it was published. Mirsky, having established the general applicability of his procedure for the extraction of nucleoproteins to a wide variety of mammalian tissues and to fish sperm, was interested in trying it out on some of our pneumococci. On July 7, I harvested a 75-liter batch of type III pneumococci, heat killed them as usual at 65°C for 30 minutes, and then turned them over to Mirsky for extraction by his procedure, with Avery and me closely following the process. Working in a cold-room at a temperature just above freezing, he proceeded to wash the pneumococcal cells three additional times with physiological salt solution and then to stir them overnight with strong salt solution in an attempt to extract the nucleoprotein. It was evident that the cells had not yielded up an amount of material comparable to that obtained from tissues, since the extract showed none of the characteristic extreme viscosity of his usual preparations. Testing a small sample by alcohol precipitation revealed that a small amount of fibrous material was present, however, and Mirsky reduced the salt concentration of the extract by dialyzing it against physiological salt solution. This resulted in the appearance of a visible precipitate, some of it in the form of fibers, which could be recovered and dissolved in a small volume of strong salt solution. Under these conditions the amount of material present was sufficient to give the typical viscosity, and he was able to reprecipitate it by his usual procedure of adding the solution to several volumes of distilled water.

Mirsky obtained a purified product by repetition of the process of redissolving and repreciptating the material and found on chemical analysis that it was made up principally of deoxyribonucleoprotein, similar to his mammalian chromosins. The yield was minuscule, however, and two further extractions of the pneumococcal cells with strong salt did not produce a significant amount of additional material. The relative inefficiency of the salt extraction was underscored when I took the residual pneumococcal cells after Mirsky was finished with them and applied our standard deoxycholate procedure. I recovered a substantial mass of fibrous DNA on alcohol precipitation of my extract—many times the amount that Mirsky had obtained—which after further purification proved to have the expected biological activity.

I tested Mirsky's nucleoprotein preparation for transforming activity, encountering some difficulties because of its solubility properties. Nevertheless, it proved to have quite respectable activity and induced transformation when present in concentrations as low as 0.2 microgram per cubic centimeter. Here, then, was a DNA-containing fraction of type III pneumococci, isolated and purified by an entirely different set of procedures, and it displayed the same biological activity. As far as I was concerned, this was a considerable boost to the idea that the transforming principle was indeed DNA, but neither Avery nor Mirsky apparently saw it that way. Admittedly this new product contained a considerable amount of protein, but I was not concerned about this, since I was confident that one could get rid of it without affecting biological activity just as we had with our own preparations. As a matter of fact, I had had to eliminate some of the protein in order to make the material soluble enough to test in the transforming system. The important point to me was that this purified DNA-containing fraction had been isolated from pneumococci by means that were totally independent from ours but was still highly active in the transformation system.

It was clear, however, that the Mirsky procedure leached too little of this material out of the pneumococci to be of any use to us in our attempts at further purification.

Not being intimately acquainted with the details of our research on the transforming principle up to this point, Mirsky appears to have attributed greater influence to this collaborative experiment on our final results than actually existed. When he and Pollister finally wrote up a complete description of their work on chromosin some four years later, they included an account of our combined pneumococcal experiment and made the following comment about it: "Because of the effectiveness of a preliminary removal of 'cytoplasmic nucleoproteins' in the isolation of the pneumococcal desoxyribose nucleoproteins by us, this procedure was subsequently used in the isolation of the transforming principle by Avery, MacLeod, and McCarty."[6] This was clearly a misconception, since it was several months prior to our collaboration that we had hit upon the device of washing our pneumococci with salt solution before extracting them with deoxycholate, and I had been using this approach as already described to reduce the amount of type III polysaccharide, ribonucleic acid, and protein in our initial extracts.

Any communication between the Avery and Mirsky laboratories had long since come to an end by the time this paper was published. I have never had a clear understanding of how and why this estrangement came about, although the passage just quoted suggests that Mirsky placed a much higher evaluation on his contribution to our work than we did. However, this never seemed to me an adequate basis for the breach or for his assuming the role, as he did, of the principal public skeptic on the subject of the DNA nature of the transforming substance. The lack of communication also caused some minor inaccuracies in his report of pneumococcal chromosin. These were not of great significance, although I did find them annoying when the paper first appeared. He reported on the transforming activity of the preparation, correctly attributing the test to me, but assigning it a potency that was ten times greater than I had actually found. I will return to consider the impact of Mirsky's views on the acceptance of our findings in discussing the aftermath of the first publication of our results.

This experiment with Mirsky that July had not deflected us from our primary purpose—to formulate a course of research that would solidify the evidence in favor of DNA. We now had in hand methods of purification that would allow us to eliminate to below the level of detectability all of the other known components of the extracts. Accordingly, we could proceed to the preparation of several lots of highly purified DNA from type III pneumococci that could be subjected to rigorous analysis by a variety of approaches. In addition to assaying our final products for transforming activity, we could apply a number of different tests to assess the composition and purity of the material. We had qualitative chemical tests that were helpful in telling us how successful we had been in getting rid of protein, carbohydrate, and ribonucleic acid. These could in turn be supplemented by the much more sensitive serological tests in the case of the antigenic proteins and polysaccharides of the pneumococcus. We could go a step further and obtain an accurate elementary analysis from the microanalyst at the Rockefeller Institute. It was also contemplated that once we had suitable preparations we would ask our physical chemist friends to look at their homogeneity in the ultracentrifuge and electrophoresis apparatus. If all of this was reasonably successful, we felt that, along with data on enzymatic inactivation, we should have the kind of evidence required.

This was not a small program. We wanted to secure adequate material for analysis and thus we planned to use the bacteria from 200 or more liters of bacteria—that is, at least three Sharples runs—for each preparation. Beginning immediately, by August I had enough washed type III pneumococcal cells stored under alcohol for us to start the first preparation. Set for the final drive, we awaited Avery's return from Maine.

IX

THE HOME STRETCH

OUR PLAN for the purification of pneumococcal DNA was nothing more than a combination of all of the steps that had proved out over the previous two years. The pneumococci were heat killed at 65°C immediately after they were harvested and then washed two or three times in physiological salt solution. We then extracted the washed cells in the usual way with deoxycholate, repeating the process as long as the extracts yielded significant amounts of fibrous precipitate on the addition of alcohol. As a matter of fact, our attention had focused sharply on these fibrous alcohol precipitates. Because of the entrapment of air bubbles as the fibers were formed during the addition of alcohol with stirring, the fibrous mass would float to the top so that it could be simply lifted out on a spatula, separating it in one step from much of the other material present in the extract. The fibrous precipitates were then redissolved in salt solution to give viscous solutions that were shaken repeatedly with chloroform and amyl alcohol by the Sevag procedure for the removal of protein.

It was at this point that we applied the treatment with the SIII enzyme, carrying out the reaction until the material no longer gave a detectable precipitate with type III antiserum. After repetition of the Sevag procedure for removal of the enzyme protein that had been added, the preparation was ready for the final stages of purification. Here we relied on the property of the DNA that permitted it to wind around the stirring rod during alcohol precipitation. The rationale was that, after the removal of the type III polysaccharide, there was no known component of the extract other than DNA that was capable of separating out in the form of fibrous strands under these conditions. Thus, by careful precipitation of the DNA at the minimal concentration of alcohol required, we should leave behind any remnants of other substances—protein, carbohydrate, or ribonucleic acid—that still remained in the preparation. When we dropped alcohol slowly into our viscous solution with constant stirring, the collection of fibers on the stirring rod began when an amount of alcohol equal to 0.8 the volume of the solution had been added and was usually complete by the time we had reached 0.9 volume. We then simply removed the precipitate on the rod, washed it in 50 percent alcohol, and redissolved it in salt solution. After repeating this process a few times, we were finished and ready to submit our product to analysis.

I don't mean to imply that things always went smoothly as we undertook to prepare several lots of this kind of material. There were a number of hitches along the way, and some modifications in the procedure were necessary as we found that we still had a few things to learn about the properties of the transforming substance. We even had some problems with simple fundamental operations, like growing the type III pneumococci for extraction. Early in the fall the central media department supplied us with a few 75-liter lots of broth that either sustained the growth of the organisms very poorly or not at all. I spent some time trying to find the source of the trouble and ended up participating in the preparation of our next batches of media. The difficulty was never pinpointed, but fortunately the difficulties disappeared with careful attention to the details of the cookbook-type recipe that had to be followed.

We were also plagued with unpredictable recurrences of "trouble with the system" leading to unsatisfactory titrations of transforming activity. On each such occasion we had to go back to the drawing board and make sure that each of the components of the system—the medium, the chest fluid, and the type II R strain—was not in some way at fault. Because of our inability to eliminate this variability entirely, we always approached the reading of one of the transforming titrations with trepidation, particularly in crucial experiments. Fess and I had a tacit agreement that neither of us would obtain a sneak preview of the results before the other had arrived in the morning. It is during this period that I have my most vivid recollection of his reaction as we converged on the incubator each day to remove the racks of tubes for reading. I can still see the expression on his face, a curious mixture of eager anticipation and apprehension for fear something had gone wrong with the complex biological system. At times like these he used one of his often repeated sayings: "Disappointment is my daily bread."

Things were not all black, however, and despite occasional setbacks due to technical difficulties we made good progress. The first of our purified preparations was ready for analysis by November and the second about a month later. Each of these had been made from the pooled organisms from three Sharples runs, representing about 200 liters of culture, and the yield of final product was less than 45 milligrams in each case. In general, these preparations lived up to our expectations. They had excellent transforming activity, and the chemical and serological tests indicated that we had been successful in our attempts to eliminate the known inactive components of the original extracts. The results of the elementary chemical analysis were also encouraging, since the values for nitrogen and phosphorus content were close to the theoretical figures calculated for pure DNA.

In getting the material ready for elementary chemical analysis, we had to eliminate the salt by dialyzing the solution thoroughly against distilled water. This led to the puzzling and unexpected finding that the purified preparations progressively and rather rapidly lost their transforming activity when retained in solution in the absence of salt, without any apparent changes in physical properties. When dissolved in physiological salt solution, the material showed no loss in biological activity after being stored for months in the refrigerator, while a distilled water solution became inactive within a few days. We were not able to explain this behavior, but once having learned about it through bitter experience we could at least avoid the problem by dialyzing only that portion of a preparation that was to be subjected to chemical analysis. Up to that point we had wasted some effort by attributing the diminished activity of our dialyzed material to the recurring troubles with the transforming system.

We encountered a second new property of the purified transforming system when we tried to keep it in the dried form. In contrast to the crude extracts, which we had been able to preserve for long periods after drying from the frozen state, the pure products lost transforming activity after drying about as rapidly as the distilled water solutions. This was true whether they were freeze-dried or dried from alcohol and ether. If a sample of the dried material were redissolved immediately after drying, the solution would have full activity; but subsequent samples on successive days showed progressively diminishing activity, with total loss within a week or so. Here again there were no detectable changes in the other properties of the pneumococcal DNA; it was still a fibrous, asbestos-like material in the dry form and gave solutions of the expected viscosity when dissolved. We were obliged, therefore, to keep the final products in salt solution and stored either in the refrigerator or, for more protracted periods, in a deep-freeze maintained with solid CO_2 (dry ice).

A large preparation, made from a total of 300 liters of pneumococcal culture, completed in mid-February 1943, and a smaller one completed a month later fully confirmed the previous findings on the quality and transforming activity of the pneumococcal DNA. Thus, when the time came to write our annual report to the Board of Scientific Directors, we were in a position to be much more definite about the progress of our attempts to determine the chemical nature of the transforming substance. Our report went to the Board in mid-April. It began with a statement on the historical background of the subject that we used later, with only minor modifications, in preparing our manuscript for the publication of this work. The tone was set by the opening sentence, which read: "Biologists, especially the geneticists, have long attempted by chemical means to induce in higher organisms predictable and specific changes which thereafter could be transmitted in series as hereditary characters." After the historical background, we described the transforming system, the current methods of purification of the active substance, and the nature of the evidence suggesting that it was DNA. There then followed a bit of interpretation and discussion of the implication of the results.

> Assuming that the sodium desoxyribonucleic acid and the active principle are one and the same substance, then the transformation from R → SIII represents a change that is chemically induced and specifically directed by a known chemical compound. Moreover, this substance selectively determines a differentiation of cellular function and structure corresponding in type to that of the S organisms from which the agent was derived. The interaction between the R cell and the transforming principle initiates a series of complex reactions which culminate in the synthesis of the Type III capsular polysaccharide. Thus, the transforming principle—a nucleic acid—and the end product of the synthesis it evokes—the Type III polysaccharide—are each chemically distinct and both are requisite in the type specific differentiation of the cell of which they form a part. The former has been likened to a gene, the latter to a gene product, the accession of which is mediated through enzymatic synthesis. The genetic interpretation of this phenomenon is supported by the fact that once transformation is induced, thereafter without further addition of the inciting agent both capsule formation and the gene-like substance are reduplicated in the daughter cells. The

changes induced are therefore not transient modifications but are transmitted through innumerable transfers in ordinary culture media.

If the present studies are confirmed and the biologically active substance isolated in highly purified form as the sodium salt of desoxyribonucleic acid actually proves to be the transforming principle, as the available evidence now suggests, then nucleic acids of this type must be regarded not merely as structurally important but as functionally active in determining the biochemical activities and specific characteristics of pneumococcal cells.

It is evident that Avery was quite willing at this juncture to be explicit about the genetic implications of our findings, at least to this audience. We have been faulted for not having been more explicit on this point in our published work. Certainly, MacLeod and I were much less inclined to be cautious, but Fess, in addition to his caution, held to the philosophy that it was enough to present the facts and leave the interpretations to others. It was not that he was indifferent to the interpretations, since he enjoyed discussing them with his associates in the laboratory and with a few other close friends, but he was more than reluctant to put his speculations in writing for public consumption. His two younger colleagues were not so inhibited but had to defer to his wishes in this matter.

Avery had passed his 65th birthday the previous October and was scheduled to be transferred to emeritus status at the end of the academic year on June 30. He had apparently not intended to continue in the laboratory after his retirement but planned to leave New York for Nashville, Tennessee, to join his brother, Roy, who had been there for some years as professor of bacteriology at Vanderbilt University. The developments in the laboratory had pretty well wiped out any enthusiasm he may have had for leaving, and he decided that he could not possibly break away until the promising work on the transforming principle had been brought to some reasonable conclusion. As far as the Rockefeller Institute was concerned there was no reason for him to leave, since it had long been the policy to provide laboratory facilities and modest support for Members Emeriti so that they could continue their research. Both of the directors, Dr. Gasser (who succeeded Dr. Flexner in 1935) and Dr. Rivers, had already urged him to stay.

On the evening of May 13, Avery sat down in his apartment to write a long letter to Roy explaining why he was not going to be able to move to Nashville that summer as originally planned.[1] He ran into some kind of block on that first evening, however, and did not finish the letter until late on the night of the 26th. A clue to what may have bothered him appears on page 3, just before the break, where he wrote: "If this War wasn't on I tell you frankly I would liquidate my affairs & start for Nashville this fall." In the interim he apparently thought better of blaming the delay solely on the war, and on resuming he launched into a detailed description of the research as a fuller and more forthright explanation for the delay. The tone of his description is less formal than in the annual report, and it is apparent that he had not talked to Roy about it previously. Everything is included: the historical background, the details of our current efforts, the speculations, and even the doubts. I did not know of this letter until more than twenty years later. Shortly after he received it, Roy had shown it to a colleague at Vanderbilt, Max Delbrück, who was well on his way to becoming the acknowledged leader of the so-called "phage group," which made remarkable contributions to biological science through the study of bacterial viruses, or bacteriophages. It was Delbrück, some time after Fess had died, who was responsible for having Roy dig the letter out of his stored memorabilia, and Roy became interested in making the letter more widely known because of his conviction that Fess had never received due recognition for his work. To my knowledge, the first publication of a portion of the letter was in 1964 in a German textbook, *Klassische und molekulare Genetik*, by C. Bresch.[2] Hotchkiss included a more extensive reproduction of the letter in an essay that he wrote in 1966 for inclusion in a volume honoring Delbrück,[3] and Dubos published the complete text—except for the introductory part that dealt only with family matters—in his scientific biography of Avery which appeared in 1976.[4]

I am guilty of having written that the letter represents the first written record of the discovery of the role of DNA as the carrier of genetic information, but it is of course antedated by the annual report, which was completed at least a month earlier. I have also noted that it served as a useful rebuttal for the view espoused by some that we were unaware of the implications of our findings. All in all, the letter is of sufficient importance to repeat part of it here. It conveys some of the special flavor of Avery's exposition under less formal conditions. We will pick it up after the historical background has been completed.

For the past 2 years, first [with] MacLeod & now with Dr. McCarty I have been trying to find out what is the chemical nature of the substance in the bacterial extract which induces this specific change. The crude extract (Type III) is full of capsular polysaccharide, C (somatic) carbohydrate, nucleoproteins, free nucleic acids of both the yeast & thymus type, lipids & other cell constituents. Try to find in that complex mixture, the active principle!! Try to isolate and chemically identify the particular substance that will by itself when brought into contact with the R cell derived from Type II cause it to elaborate Type III capsular polysaccharide, & to acquire all the aristocratic distinctions of the same specific type of cells as that from which the extract was prepared! Some job—full of heartaches & heart-breaks. But at last *perhaps* we have it. The active substance is *not* digested by crystalline trypsin or chymotrypsin—It does not lose activity when treated with crystalline Ribonuclease which specifically breaks down yeast nucleic acid. The Type III capsular polysaccharide can be removed by digestion with the specific Type III enzyme without loss of transforming activity of a potent extract. The lipids can be extracted from such extracts by alcohol & ether at −12°C without impairing biological activity. The extract can be de-proteinized by the Sevag Method (shaking [with] chloroform & amyl alcohol) until protein free and biuret negative. When extracts, treated & purified to this extent, but still containing traces of protein, lots of C carbohydrate & nucleic acids of both the yeast

& thymus types are further fractionated by the dropwise addition of absolute ethyl alcohol, an interesting thing occurs. When alcohol reaches a concentration of about 9/10 volume there separates out a fibrous substance which on stirring the mixture wraps itself around the glass rod like thread on a spool—& the other impurities stay behind as granular precipitate. The fibrous material is redissolved & the process repeated several times—In short, this substance is highly reactive & on elementary analysis conforms *very* closely to the theoretical values of pure *desoxyribose nucleic acid* (thymus type). Who could have guessed it? This type of nucleic acid has not to my knowledge been recognized in pneumococcus before—though it has been found in other bacteria.

Of a number of crude enzyme preparations from rabbit bone, swine kidney, dog intestinal mucosa, & *pneumococci,* and fresh blood serum of human, dog, and rabbit, only those containing active depolymerase capable of breaking down known authentic samples of desoxyribose nucleic acid have been found to destroy the activity of our substance—indirect evidence but suggestive that the transforming principle as isolated may belong to this class of chemical substance. We have isolated highly purified substance of which as little as 0.02 of a *micro*gram is active in inducing transformation. In the reaction mixture (culture medium) this represents a dilution of 1 part in a hundred million—potent stuff that—& highly specific. This does not leave much room for impurities—but the evidence is not good enough yet. In dilution of 1:1000 the substance is highly *viscous* as an authentic preparation of desoxyribose nucleic acid derived from fish sperm. Preliminary studies with the ultracentrifuge indicate a molecular weight of approximately 500,000—a highly polymerized substance.

We are now planning to prepare new batch & get further evidence of purity & homogeneity by use of ultracentrifuge & electrophoresis. This will keep me here for a while longer. If things go well I hope to go up to Deer Isle, rest awhile—Come back refreshed & try to pick up loose ends & write up the work. If we are right, & of course that's not yet proven, then it means that nucleic acids are *not* merely structurally important but functionally active substances in determing the biochemical activities and specific characteristics of cells—& that by means of a known chemical substance it is possible to induce *predictable* and *hereditary* changes in cells. This is something that has long been the dream of geneticists. The mutations they induce by X-ray and ultraviolet are always unpredictable, random, & chance changes. If we're proven to be right—and of course that's a big *if*—then it means that both the chemical nature of the *inducing stimulus* is known & the chemical structure of the *substance produced* is also known—the former being thymus nucleic acid—the latter Type III polysaccharide and both are thereafter reduplicated in the daughter cells. And after innumerable transfers and without further addition of the inducing agent, the same active & specific transforming substance can be recovered far in excess of the amount originally used to induce the reaction. Sounds like a virus—may be a gene. But with mechanisms I am not now concerned—One step at a time—& the first is, what is the chemical nature of the transforming principle? Someone else can work out the rest. Of course, the problem bristles with implications. It touches the biochemistry of the thymus type of nucleic acids which are known to constitute the major part of the chromosomes but have been thought to be alike regardless of origin & species. It touches genetics, enzyme chemistry, cell metabolism & carbohydrate synthesis, etc. But today it takes a lot of well documented evidence to convince anyone that the sodium salt of desoxyribose nucleic acid, protein free, could possibly be endowed with such biologically active & specific properties & this evidence we are now trying to get. It's a lot of fun to blow bubbles—but it's wiser to prick them yourself before someone else tries to.

So there's the story Roy—right or wrong it's been good fun & lots of work. This supplemented by war work and general supervision of other important problems in the Lab has kept me busy, as you can well understand. Talk it over with Goodpasture*but dont shout it around—until we're quite sure or at least as sure as present method permits. It's hazardous to go off half cocked—& embarrassing to have to retract later.

In addition to its historical implications, this letter gives a pretty good picture of the state of affairs in the laboratory in May 1943. The "new batch" to which Avery referred was already under way, having been started at the time the letter was being written. The combined organisms from two 75-liter lots were used in a slightly modified method which consisted of carrying out all of the procedures in the cold except those that required higher temperature, like extraction with deoxy-

* Dr. Earnest Goodpasture, a professor at Vanderbilt and a good friend of both Averys.

cholate and treatment with the SIII enzyme. Everything went smoothly and by the end of the first week in June we had completed the chemical and serological analyses that showed the new preparation to be up to the standards of our previous lots. Perhaps, not surprisingly, in view of the special care that we had exercised in keeping the material cold throughout the purification process, it was unusually high in biological activity. As little as 0.003 microgram (3×10^{-9} gram) of the final product was effective in producing transformation in 50 percent of the tubes. This was the preparation with which we then proceeded to "get further evidence of purity and homogeneity by use of the ultracentrifuge and electrophoresis."

Dr. Rothen made several ultracentrifuge runs first to determine whether the optically visible boundary seen during centrifugation coincided with the biological activity and second to confirm his earlier estimate of the molecular weight. Fortunately the quantitative titration of transforming activity was working beautifully at this time so that we could accurately measure the amount of T.P. in the fractions removed from the centrifuge cell. The results were most encouraging. The purified material had a single, exceedingly sharp boundary as it sedimented in the ultracentrifugal field, and the transforming activity clearly moved with this boundary. We had similar good news from the electrophoretic studies carried out by another physical chemist colleague, Theodore Shedlovsky. From the application of optical techniques like those used with the ultracentrifuge, it was shown that there was only one moving boundary as the material moved in an

ON THE FACING PAGE: A page from the laboratory notebook recording a test of transforming activity of the final preparation before writing up the work. The + marks in the table refer to the presence of diffuse growth, presumptive evidence for transformation. SIII indicates that type III pneumococci were recovered after plating out the cultures for confirmation. R indicates that only rough organisms were recovered. Not all of my laboratory notes were this tidy, although the principle of preparing thorough, clearly legible records was adhered to throughout the research.

6/7/43

Transforming Activity - T.P. #43-44 ✓

T.P. #43-44 - Final purified product. 2.0 cc. pptd by alcohol 6/5/43
Solution contained 0.52 mg/cc.
Taken up in 2.0 cc. neut. saline. Half-log dilns in neut. saline

C.A. Neo V + 10% Rockwell chest fluid - (5/18/43. Heated 60°/30' 5/25/43)

R36A - 5½ hour growth in C.A. Neo V blood broth. 10^{-4} diln 0.05cc inoc.

0.2 cc T.P. dilutions ⟶ quadruplicate tubes

Tube	Dilution	Amt. added (milligrams)	- a -	- b -	- c -	- d -
1	1:100	0.001	+ S III	+ S III	+ S III	+ S III
2	1:300		+ S III	+ S III	+ S III	+ S III
3	1:1000	0.0001	+ S III	+ S III	+ S III	+ S III
4	1:3000		+ S III	+ S III	+ S III	+ S III
5	1:10,000	0.00001	+ S III	+ S III	+ S III	+ S III
6	1:30,000	0.000003	− R	+ S III	− R	+ S III
7	1:100,000		− R	− R	− R	− R

Comment

This is a highly active preparation which is active in amounts as small as 0.003 micrograms (or 3×10^{-9} grams). In the 2cc reacting system, this represents a dilution of 1 part in 600,000,000.

μ gms.

electric field and that again this was associated with the biological activity. Thus we had the added assurance that not only was our purified pneumococcal DNA homogeneous by two quite different physical tests but also that it was extremely unlikely that its transforming activity resided in some other, undetected component.

Avery's doubts were still not altogether resolved. How could we establish with certainty that transformation was not attributable to some unknown substance that was carried along in our purification process and remained as a minor constituent of the final product? This question led us to make a pilgrimage to the Princeton laboratories of the Rockefeller Institute to consult with two of our colleagues there, John Northrop and Wendell Stanley. Both Northrop and Stanley were to win the Nobel Prize in 1946, Northrop for his crystallization of pepsin, the protein-splitting enzyme of the stomach, and Stanley for crystallization of the tobacco mosaic virus. The work of both men had been greeted with much skepticism when first reported. The principal question was how did they know that their crystals were not merely carrying along a small amount of a contaminant that possessed the biological activity. It seemed to us, therefore, that they had faced problems similar to ours.

We had seen less of Colin MacLeod during this period because, in addition to his task of building up and running the department of microbiology at New York University, he had become heavily involved in activities related to the war effort, serving as consultant to the secretary of war and director of the Commission on Pneumonia of the Army Epidemiological Board. He found time, however, to join Fess and me in our trip to Princeton. Northrop and Stanley were certainly understanding and sympathetic, but they had no magic formula for solving our problem and no specific procedures to suggest. They seemed impressed with the data we had in hand, and their only advice was that we should marshal all of the evidence that we were able to obtain and proceed with publication.

Even though the conference had ended on a distinctly upbeat note, Fess remained hesitant. I can remember that as we discussed the situation on the way home on the train, Colin asked him with a certain amount of impatience: "What else do you want, Fess? What more evidence can we get?" I don't believe that he replied to this, but one answer that he had was to seek still more advice. I can recall the session that Fess and I had for this purpose with Van Slyke, a good friend and colleague who had been at Rockefeller since 1907 and who was a distinguished biochemist. He also offered no specific advice, giving us the same sort of encouragement we had gotten at Princeton. Strangely, in another manifestation of those memory gaps, I had completely forgotten a similar meeting with Max Bergmann, another biochemist who had come to the Institute from Germany in 1934. He had been one of the early workers to be concerned with the structure of proteins, and he developed at Rockefeller a remarkable research group in this field. The evidence that this meeting actually took place is a loose yellow sheet that I found among my notes on which I had pencilled what was apparently a summary of his comments in reply to our queries. It is unfortunately undated but almost certainly relates to this same period, since Bergmann died in 1944. What makes the lapse in memory all the more inexplicable is the nature of his comments, which are more emphatically supportive than anything we had heard from others. I headed my notes, "Interview with Dr. Bergmann," and began with the following paragraph:

> In the light of present knowledge, the statement that all nucleic acids are the same regardless of the source from which they are derived is nonsense. If they are large polymeric compounds, there is an endless number of possible combinations all of which would possess the same elementary composition but would differ in chemical structure none the less. Nucleic acids hold too prominent a place in biology to be completely non-specific substances. The lack of evidence of any specificity associated with nucleic acids is only due to the fact that they have not been investigated sufficiently.

This is rather strong stuff for that particular period in the history of nucleic acids, and I can't be sure how faithfully I had paraphrased his comments without injecting some interpretations of my own. I go on to a list of several suggestions that Bergmann made for obtaining further evidence. Some of these we had already tried, such as salt fractionation of the extracts, and others (e.g., chromatography) we were not up to tackling with the methodology then available. In any event, my notes suggest that he was more responsive than the others we consulted, and I have no explantion for the obliteration of this episode from my memory.

I do not believe that Fess was greatly reassured by these consultations, but in the end he yielded and agreed that it was probably time to begin writing up the work. He would go off to Maine for the summer and give it some further thought, but at the same time would prepare drafts of the introduction and the discussion. It was my job to write up the methods section of the paper and to collect the experimental protocols that we would use to document the results. I also had a few odds and ends of experimental work to do in the laboratory in order to tidy up some of the details. One of these had to do with the statement about the transforming substance being recoverable from the transformed cells in amounts far in excess of that originally used to induce the transformation. In other words, the R strain, after being induced to form the type III capsular polysaccharide, continued to reproduce the inducing agent as well as the polysaccharide. It seemed certain that this must be so, but as far as I could tell no one had ever demonstrated it experimentally—that is, isolated a colony of the transformed cells, grown them up in a few liters of culture, and shown that one could extract as much transforming principle as one could from the original type III strain. I did the experiment in July with the expected result—the transformed cells were an excellent source of the active material—and wrote Fess about it along with other bits of new information that he might want to use in writing the discussion.

When Fess returned in the fall we got down to the job of preparing the paper in earnest. In order to avoid the interruptions of the telephone and laboratory activity, we obtained a small room in the library and were cloistered together there for a few hours each day. We began by going over the initial drafts that each of us had prepared and trying to decide what else should be included. I remember being so bold as to suggest that we use some information from the experiment that we had done with Mirsky, since as I said before I considered it to be supportive of our thesis, but Fess quickly vetoed this idea. He was probably right, because it was a single experiment with such a minute yield of active material that complete analysis was not possible. He may have had other motives besides his view of the adequacy of the data and their bearing on our conclusions, but I was not aware at the time that there had yet been any cooling of the relationship between the two laboratories.

The writing went slowly, as the casting of every phrase had to be carefully scrutinized in the manner for which Fess was famous. It turned out to be a rather long paper—twenty printed pages—necessarily including a description of the results of earlier work, largely by MacLeod, which had established a more reliable transforming system, as well as examples of all the recent data. As we neared completion of the final draft in October, we suddenly realized the value of a photograph showing the dramatic difference between colonies of the R strain and those of the transformed type III organisms. We had not bothered to stop along the way to get illustrative material of this kind, and I had to scramble to have suitable photographs taken in a hurry. On October 28, just three days before we submitted the paper for publication in the *Journal of Experimental Medicine*, I spent a few hours with the photographer in the illustration division of the Institute shooting pictures of a blood agar plate with colonies of the inoculating R strain, R36A, on one half and colonies of transformed type III cells on the other. Finally, by trying a variety of lighting

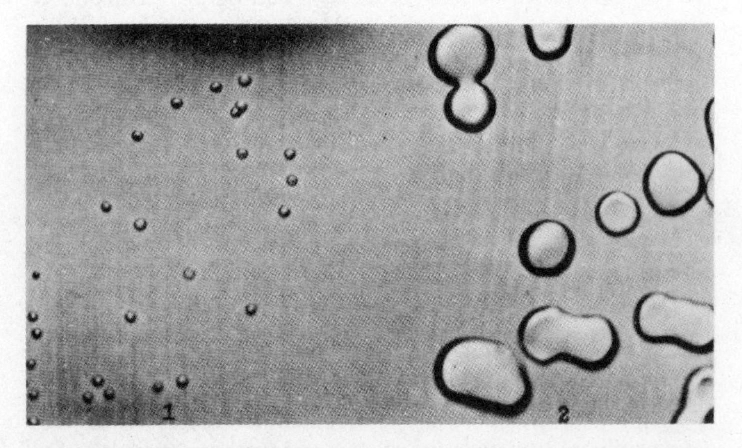

Colonies of pneumococci on the surface of blood agar. This is the picture used in the 1944 paper. The small colonies on the left are of R36A, the unencapsulated strain derived from type II. The smooth, glistening colonies on the right are of the same strain after transformation with DNA from type III. They have capsules of type III polysaccharide. *(Reproduced from* The Journal of Experimental Medicine, *1944, 79:137–158, plate 1, by copyright permission of The Rockefeller University Press.)*

angles and cropping out areas that were out of focus, we came up with an acceptable picture. By contrasting the small, rough-surfaced R colonies with the much larger, juicy-appearing colonies of organisms that had been transformed to produce type III capsules, it showed the magnitude of the change and turned out to be a useful addition to the paper. It has since been reprinted in many books to illustrate the phenomenon of transformation.

An amusing episode occurred during this period when Fess discussed with me his concerns about the order in which our names should appear on the paper, a matter that causes more trouble among scientists than the layman might imagine. He said that he wasn't sure whether the names should appear in the order of the length of association with the problem, on the basis of age and seniority, or simply alphabetically. It was not until after he had left me on that occasion that it suddenly hit me that all of the alternatives came to the same result. No matter how you sliced it, it was "Avery, MacLeod, and McCarty." It was fine with me.

After the manuscript was put in final form and thoroughly proofread, Avery delivered it by hand on November 1 to the editor of the *Journal of Experimental Medicine,* Peyton Rous. Rous was another longtime colleague, having joined the staff of the Rockefeller Institute four years before Avery. He was a gifted and versatile experimentalist in his own right (he found in 1910, for example, the first virus that causes cancer in animals, a discovery for which he was awarded the Nobel Prize more than fifty years later), and at the same time labored for years building and sustaining the reputation of the *Journal.* He was a skilled and strict editor, demanding clarity and correct English as well as scientific accuracy from his contributors. Avery told his old friend on submitting the manuscript that he wanted him to go over it just as he would if it were a paper submitted by an unknown outsider, and Rous took him at his word.

Avery had me join him in his office when Rous personally

delivered the edited manuscript to discuss his suggested changes. He began by reminding Fess that he had been asked to give it the full treatment and proceeded to bring up a long series of points. The typescript was covered with many lightly pencilled notations in Rous's fine handwriting. Most of these dealt with minor queries or suggested changes in wording, but there were a few more substantive comments. We had included in the discussion that quotation from the paper by Leathes in which he stated that "it becomes a question whether the virtues of the nucleic acids may not rival those of the amino acid chains in their importance," and Rous pointed out that this was pure speculation and really helped very little to support our thesis. In the end, it was deleted. I have a vivid recollection of his reaction to my use of the word "routinely" at two or three places in describing the methods. He called attention to the unacceptability of the word in a colorfully phrased note—"Saving your presence, 'routinely' is a louse on the dictionary." I was amazed to find that he was correct and that none of the standard dictionaries had yet caught up with the fact that this adverbial form was in wide use. The word was replaced with the phrase "as a routine" in the corrected version of the paper.

I should not give the impression that our discussions with Rous assumed a contentious quality, because all of the points at issue were resolved amicably and rather quickly. It is a pity that no copy of this edited manuscript still exists to illustrate the Rous style and to remind me of some of the details of the editing that I am sure I have forgotten. As a matter of fact, due to Avery's rather ruthless cleaning out of his files on his departure from Rockefeller in 1948, there is not even a copy of the revised version of the paper. I cannot recall that Rous at any time indicated what he thought about the work or how he viewed its broader implications, but some indication of the latter emerged after his death in 1970. Dr. Rous's extensive collection of reprints of scientific articles had come into the hands of Dr. Paul Cranefield at Rockefeller who discovered that Rous had written on his copy of a reprint of our paper, as instructions to his secretary: "Please file under genetics."

The paper was ready for the printer by December, and all we had to do was wait, somewhat impatiently, for it to appear in the *Journal*. I was not immune to concerns about having some of my experimental work published if I were to succeed in establishing myself in a research career—a manifestation of the "publish or perish" syndrome—and I had chafed a little under the onus of having worked for well over two years without a publication. Accordingly, my impatience was fed by this attitude, as well as by the more substantial consideration based on my conviction that our discovery had broad biological significance and the sooner it appeared in print the better. There was, however, plenty to do in the laboratory to keep us busy in the meantime.

There was also the matter of reporting on our findings to our colleagues in the other laboratories of the Institute. We did not have the profusion of lectures, seminars, and other meetings that exist today, but every Friday afternoon there was a staff meeting, attended by essentially everyone, at which recent work from one of the laboratories was presented. It was an effective mechanism for keeping us up-to-date on the activities of the other research groups at the Institute. Fess had not presented any of the work from his laboratory before this forum for many years, and Colin and I felt that he should tell the transformation story there, now that we had the paper in press. After initially expressing some reluctance, he finally agreed that it was appropriate and the talk was scheduled for December 10, 1943. It is listed in the records of the staff meetings under the same title as the paper: "Studies on the Chemical Nature of the Substance Inducing Transformation of Pneumococcal Types," with all three of our names.

The room in which these meetings were held was not large but was adequate for the size of the staff, although on this occasion there were enough visitors from the outside so that a few of those attending had to stand at the back of the room.

Avery's talk followed the same sequence and much of the phraseology of the paper, and it closed, as I remember it, with the same conclusion that we had used to end the paper: "The evidence presented supports the belief that a nucleic acid of the desoxyribose type is the fundamental unit of the transforming principle of Pneumococcus Type III." He received a resounding round of applause, but when the chairman called for the customary questions or discussion, there was no response. The regular chairman of the staff meetings that year, Frank Horsfall, was absent because of illness, and the secretary, Howard A. Schneider, was in the chair. He later recalled the situation as follows: "At the conclusion of his polished and long-worked-over address, I rose to call for discussions and questions. No one rose to my call. No one spoke. There followed one of those long silences that haunts me yet. Instinctively I felt we were witnesses to something important, even though I cannot say I fully appreciated just *how* important that paper was to become as the years unrolled. And then, one man arose. It was Dr. Michael Heidelberger, an old colleague of Fess and then professor of immunology at Columbia. I cannot remember all he said, although Dr. Heidelberger was brief in his remarks. I do recall his describing the years of thought and incubation that he personally knew, as a former colleague, lay behind the afternoon's paper. When Dr. Heidelberger sat down another long silence ensued. At last, when I could stand it no longer, I said, 'This company, having reached an unanimity of opinion, is now adjourned.' "[5]

My recollection is not too different from Schneider's, only I don't believe that I considered the silence embarrassing or strange. The applause had reflected the esteem and affection which most of the staff held for Fess almost as much as it did a response to the content of the talk. No one was prepared to comment on the latter on the spur of the moment. A few have mistakenly recalled that after the lecture Mirsky had come to the podium at the front of the room and delivered his first public rebuttal, indicating why he thought it was going beyond the evidence to conclude that the transforming substance was DNA. This actually occurred much later in a quite different situation. There were no expressions of doubt on the occasion of Fess's valedictory.

The *Journal of Experimental Medicine* prided itself in appearing promptly on the first of each month. On February 1, 1944, the issue bearing our paper was on the library table with the other new accessions for the day.[6] This marked the end of what seemed to me to have been a very long search. Recently my brother returned to me the copy of the reprint of the paper that I had sent to my mother as soon as they were available, and I find that I had expressed this sentiment by writing on the reprint: "This is it, at long last."

X

STRENGTHENING THE EVIDENCE

EVEN THOUGH we looked upon the publication of the paper as a milestone, we did not by any means consider that it marked the end of the job. We had already heard enough about nucleic acids being all alike to know that there were going to be doubts about our findings. The question was: how could we get additional, compelling evidence that it was the DNA itself and not some protein remaining in our purified preparations that was responsible for the biological activity?

I was not very optimistic about getting this kind of evidence by further purification of our transforming DNA. My lack of confidence in this approach was based on a bit of chemical calculation. There is a physical constant called Avogadro's number which states that a gram molecule of any substance—that is, an amount equal to the molecular weight in grams—contains 6.022×10^{23} (or roughly 6 followed by 23 zeros) molecules, manifestly a very large number. This would apply as well to large molecules and, if we assumed a molecular weight of 1 million for our DNA, it would have that number of molecules in a million grams. However, even 1 milligram would have 6×10^{14} molecules, and the smallest amount of our best preparation that was able to induce transformation—0.003 microgram (3×10^{-9} gram)—would still have more than a billion. This apparent excess didn't bother me, since we had no idea how the DNA managed to get into the R pneumococcus and become integrated into the cell's mechanisms in order to initiate the production of type III capsular polysaccharide, and it seemed unlikely that this could be a very efficient operation. Furthermore, if we were right, only a small portion of the total DNA in our preparations could be specific for capsular synthesis and the rest of it had to be concerned with various other functions of the pneumococcal cell. One could assume that there had to be an excess of DNA molecules in the transforming system.

The implications of these numbers with respect to protein contamination were more troublesome, however. If we made the optimistic assumption that there was no more than 0.1 percent protein in our final product, the minimal transforming dose of 0.003 microgram would still have millions of protein molecules in it. The available analytical methods were not up to detecting contaminating protein at this level, and I could see no prospect of devising methods of purification that would assure us of having eliminated essentially all of the protein. We already knew that the hypothetical protein would have to have rather special properties, since it resisted our deproteinization procedures, was unaffected by several proteolytic enzymes, and was carried along without loss through all the steps in the purification of the DNA. We concluded that we should look for an alternative approach in attempting to validate the identification of the transforming substance as DNA. The idea that appealed to me, and one that had certainly been fostered by my experience with Avery up to that point, was to get a purified and well-characterized enzyme that could digest DNA and to show that it would destroy the transforming substance. This was pretty much in line with Avery's thinking back in the 1920s when he was confronted with the necessity of convincing skeptics that his polysaccharides themselves possessed serological specificity rather than being dependent on contaminating protein. We needed something like the SIII enzyme that he and Dubos had finally come up with at that time. The enzymatic experiments included in our paper, showing that all of the crude preparations capable of inactivating the transforming principle also acted on purified mammalian DNA, were all right as far as they went, but they were only correlations and thus not conclusive. What we needed was a well-characterized deoxyribonuclease that would be generally effective in splitting DNA in much the same way that the crystalline proteolytic enzymes already known at that time were able to attack most proteins.

I got very little help from a search of the biochemical literature. No one appeared to have made a serious effort to obtain even a partially purified deoxyribonuclease. This was another reflection of the rather poorly developed state of nucleic acid biochemistry at the time this work was in progress. Kunitz had reported his isolation of crystalline ribonuclease in 1940, and so this was also a relatively new product for application to biological studies. About the most useful information that I could glean from the earlier literature on DNA-splitting enzymes was that the mammalian pancreas, the same source used by Kunitz for his ribonuclease, was likely to be the richest source of the DNA enzyme. Otherwise, there was not much to be found about its properties or methods of preparation.

Our interest in having a bona fide deoxyribonuclease had not waited for the completion of the first phase of the work and publication of the paper. As early as the summer of 1942, after we had accumulated the various kinds of evidence pointing to DNA as the transforming substance, we realized that the availability of such an enzyme would be of great help. Our little chemical laboratory was presided over by a biochemical technician, William La Rosa, whom Avery had more or less inherited from Alexis Carrel at the time that the latter had closed his laboratory in 1940. La Rosa had been engaged in some work related to the sulfonamide drugs with Colin MacLeod and now seemed available for tackling exploratory experiments on deoxyribonuclease. (Our laboratory shorthand for the enzyme was for some years "dornase," but the generally accepted abbreviation today is DNase, and for ribonuclease, RNase.)

La Rosa was a good chemist, but he had been given a measure of independence in his previous work under Carrel that made him resistant to carrying out faithfully the research protocols that I drew up for our attempts to purify DNase. He tended to inject his own ideas into the research and rarely managed to complete an experiment as I had outlined it. I found this frustrating since the results never satisfied me, but I was too heavily engaged in the work on the transforming factor to carry out the experiments myself. As a result, there was little progress toward obtaining a DNase during this period. One important piece of information did come out of his efforts, however, when he found that the enzyme present in pancreatic extracts seemed to require the presence of magnesium ion (Mg^{2+}) in order to exert its activity on DNA. This was in contrast to Kunitz's RNase, which had no such requirement for metal ions, but I was able later to confirm La Rosa's observations fully and to make some practical applications of this information in the study of the enzyme's effect on transforming DNA.

La Rosa left Rockefeller by the summer of 1943, freeing the chemical laboratory for me to try my own hand at the DNase experiments whenever there was a lull in the activities relating to the preparation of the paper on the DNA nature of the transforming substance. I used the same starting material that I had had La Rosa use—a commercial preparation of a dried extract of beef pancreas. This brownish powder was far from being ideal for the purposes of chemical fractionation, but it had the advantage of being readily available and it did have potent DNase activity. Measurement of enzyme activity in the various fractions was something of a headache, since no simple chemical method was available. The most obvious physical effect of the enzyme on a solution of DNA of the Mirsky type was the loss of its characteristic viscosity, accompanied by a loss of its ability to give fibrous alcohol precipitates, and I chose to measure the rate of fall of viscosity as an index of DNase activity. This meant that the reaction had to be carried out in a viscosimeter, with repeated readings being taken over a period of time in order to determine the rate of change. It was a cumbersome method, but at least it was sensitive and also reliable in telling us what we wanted to know—that is, it could be shown that the rate of fall of viscosity was proportional to the concentration of enzyme present.

By applying standard salt fractionation procedures, I was able to separate material from the pancreatin that had far more DNase activity than the crude enzymes and sera that we had used in making the experimental correlations we were about to publish. These products were themselves still very crude but good enough to permit initial studies on the optimal conditions for measuring the enzyme's activity. Even at this early stage I couldn't resist trying their effect on some of our transforming DNA, and the results gave us some indication, even before the paper was published, that a purified DNase was going to back up our story. The data were too preliminary and the purification attempts had not progressed far enough, however, for us to incorporate any of this information into the paper.

Early in January 1944 we acquired a replacement for La Rosa as a technician in the chemistry laboratory to help me with these studies. Up to the onset of the war the technical staff of the Rockefeller Institute had been nearly all male (in our group there were La Rosa, Fred Kimmer, and Teddy Nadeje), most of whom had come as youngsters and grown up with the place. The war changed the situation rapidly, as all of the replacements and the new technicians hired for the special projects were young women, and they were in the majority by the end of the war. Our representative of this new wave was Jacqueline Jonkowske, who had previously been working in a hospital clinical laboratory and was happy to escape routine analyses by moving to a research lab. She was ideal for the job: technically competent, able to follow directions, and with enough patience to sit for hours in front of viscosimeters measuring the enzyme activity of different fractions.

As it turned out, I needed her help even more than I had suspected. During that winter I began to develop the symptoms of arthritis, manifested at the outset principally by morning stiffness. I can still recall the agony of trying to hobble downstairs on awakening to set up the thermostat and take some of the chill off our underheated house—a victim of oil rationing. My feet then began to bother me, and I found it difficult to stay on them all day in the laboratory as I had been accustomed to do. Subsequently, as other joints became inflamed and swollen, my effectiveness in the laboratory was clearly suffering, and in the middle of May I tossed in the towel and entered our hospital as a patient. Whether or not it was therapeutically wise to have my hospital bed in the same building as my laboratory, it made it possible for me to keep some of the work going. As far as therapy was concerned, I was fortunate to be under the care of Dr. Robert F. Watson, who had been the resident physician of the hospital when I arrived in 1941 and had continued in a similar capacity after the unit was called to active duty. Watson was associated with the rheumatic fever service and thus had some expertise in dealing with the problems of joint disease, but, more important, he was a superb and judicious physician.

I had seen enough of the ravages of rheumatoid arthritis to dread the possibility that I was suffering from it. When the Normandy D-Day came along after I had been in the hospital for three weeks, my morale was at a low ebb. Two of the leading rheumatologists in New York City had been brought to see me in consultation, and they agreed that rheumatoid arthritis was the most likely diagnosis. Both recommended that I be started on gold therapy. Fortunately, Watson wasn't any more enthusiastic about this recommendation than I was and, with the support of a consultation by mail with a third specialist, he held off any specific treatment while waiting to see what further developments there were on simple bed rest. To everyone's surprise, I began to show definite improvement and was able to return home on July 7 with only a few residual symptoms, which proceeded to disappear permanently over the next few weeks. Had I received the gold treatment, my case would have gone on the record as a remarkable cure for that mode of therapy.

The laboratory notes suggest that, despite my illusions about keeping the work going, not very much had really gone on during this period. In February we had given up the use of dried pancreatin as the source of the enzyme and had turned to extraction of fresh beef pancreas, which Fred Kimmer was able to pick up for us as we needed it by stopping off at the slaughterhouse on his way to work. Nevertheless, we continued to have trouble in reproducibly separating out clean fractions of the active enzyme, and we were troubled in addition by the ability of the protein-splitting enzymes, which abounded in the pancreatic extracts, to destroy the DNase. Our efforts to circumvent these problems continued along with experiments designed to define better the properties of the enzyme and the conditions for its accurate measurement.

Things were still in this state when I returned to the laboratory in August and decided that it was high time to consult Kunitz about the approaches that he used in fractionating his pancreatic extracts. He had not only obtained RNase from this source but together with Northrop had earlier isolated crystalline preparations of the two major protein-splitting enzymes, trypsin and chymotrypsin, and of their inactive precursors from this same starting material. It is difficult to imagine now why I was so slow about seeking his help, since I was certainly aware of his work. It would seem that the search for a DNase had too long been merely a sideline, relegated to a second priority until the work on the purification of transforming DNA was completed. When I called Kunitz, he suggested that I come to his laboratory and observe his techniques. I quickly arranged for my second visit to the Princeton laboratories of the Rockefeller Institute.

Kunitz was an old-timer, having come to Rockefeller the same year as Avery. He had initially held a technical position but then gradually rose through the ranks of the scientific staff until he finally became a full Member in 1949. In the course of his association with Northrop in their early work on proteins, he turned out to be an absolute genius at the business of crystallizing proteins, and he tends to be remembered primarily for this even though he also carried out extensive studies of the properties and biological significance of the proteins that he isolated. He was a delighful little fellow, about as tall as Avery, and with a heavy accent that he had carried over from his native Russian. He could not have been more hospitable or more generous with his time in helping me with my problem.

The Kunitz laboratory was beautifully equipped for the large-scale preparation of enzymes from various sources, and the standard procedures that he had devised were elaborately worked out and systemized. The net effect was to make his published methods highly reproducible if one faithfully followed his directions, as I found out later in repeating some of his preparations. The heart of his fractionation process was the application of the time-honored technique of "salting out" proteins, usually with ammonium sulfate. Ammonium sulfate has a number of advantages for this kind of work, one of the most important being its high solubility—it takes 760 grams (about 1.67 pounds) of the salt to fully saturate 1 liter of water. Kunitz had constructed a formula for calculating the amount of ammonium sulfate required to bring a protein solution to any desired level of saturation, or to bring it from one level to another, and he had prepared tables of the data for quick application in the laboratory. These data, together with hints on filtration methods and a few other tricks of the trade, were to be of great help to me. The key to his success with pancreas was the use of a dilute sulfuric acid (about 2.5 percent) instead of water in the initial extraction of the ground organ. The acid inhibited the activation of the proteolytic enzymes and at the same time eliminated a problem with mucoid substances that interfered with fractionation and filtration.

It did not take long after I returned to New York to get together the necessary materials to apply the Kunitz technique to the isolation of DNase. The first step in his procedure, after obtaining the acid extract of the pancreas, was to bring it to 0.4 saturation with ammonium sulfate and to discard the precipitate that was formed, since all of the several enzymes in which he and Northrop were interested remained in the filtrate at this point. My first experiments with his procedure revealed that the DNase was wholly contained in the rejected 0.4-saturated precipitate. It was painful to realize that during the processing of vast quantities of pancreas in the Princeton laboratories over the previous years all of the DNase had been tossed out with the garbage. This observation had its compensations, however, since it suggested that it should not be too difficult to separate the DNase from the bulk of the trypsin, chymotrypsin, and ribonuclease, as well as the other enzymes with which Northrop and Kunitz had been concerned.

I modified the Kunitz procedure by bringing the acid pancreatic extract to only 0.2 saturation with ammonium sulfate before filtering to obtain a clear solution and then increasing the concentration of the salt to 0.4 saturation in order to precipitate the DNase. Very quickly I had preparations that were much more potent than anything I had made previously, and with some reworking of this fraction I obtained material that seemed most promising for the purposes that Avery and I had in mind. The purified product would cause a readily detectable fall in the viscosity of a DNA solution when used in concentrations as low as 0.01 microgram per cubic centimeter; and it took 100,000 times as much material to demonstrate any protein-splitting activity. When we tried it out on our pneumococcal DNA preparations, the same low concentrations totally destroyed transforming activity, and with even smaller amounts of the enzyme (0.001 μg/cc) the activity was clearly diminished. The enzymatic approach to verification of the DNA nature of the transforming substance was living up to our expectations.

We got some additional encouragement as a result of further studies on the activation of DNase with metal ions. We found first that, among a number of other metal ions, only manganese ion (Mn^{2+}) was as effective as magnesium ion (Mg^{2+}) in activating the enzyme. It had occurred to me that sodium citrate (a salt of citric acid, the principal organic acid of citrus fruits) should inhibit the magesium-activated DNase by virtue of its well-known ability to tie up Mg^{2+} in a complex so that it was no longer free. This turned out to be the case, and citrate eliminated completely the fall in viscosity of a DNA solution exposed to the Mg^{2+}-activated enzyme but had no inhibitory effect whatever when Mn^{2+} was used as the activator. The same pattern held in experiments with pneumococcal DNA: the destruction of transforming activity by Mg^{2+}-activated DNase was prevented by the presence of citrate but the action of the Mn^{2+}-activated enzyme was unaffected. This made it all the more certain that it was the DNase itself that was acting on the transforming substance.

I made a number of attempts to crystallize the DNase, since this seemed about the only feasible means of achieving further purification to get rid of the residual proteolytic activity and other probable contaminants in my preparations. The powerful modern methods for the separation and purification of enzyme proteins were still a long way off. I had saved all of the residues after removing the DNase from my pancreatic extracts and I was thus able to use this material to get experience with protein crystallization by repeating the experiments of Northrop and Kunitz. This is how I discovered the precision and reproducibility of their published methods.

Before long I had my own crystalline preparations of five of their enzymes, but none of this experience was translated into any success in crystallizing DNase. Kunitz is reported to have remarked during a lecture that for success in work of this kind "All you need is a barrel of ammonium sulphate and a drum of concentrated sulphuric acid." Northrop added in discussion, "One also needs a barrel of patience."[1] This, or some more subtle ingredient, is what I seemed to lack.

While this work was in progress an event occurred that revealed something of how the younger members of the hospital staff viewed the work on pneumococcal transformation. This was at a dinner of the staff, held at the Harvard Club on November 2, 1944, that had a dual purpose: to give a send-off for that part of our naval research unit that was about to depart for the southwest Pacific and to honor Fess after his recent retirement. (I made the error in an earlier publication[2] of giving the date of this dinner as April 1943, even though I was vaguely aware that the southwest Pacific theme came at the same time as the homage to Avery.) Part of the entertainment at the dinner was a series of verses written by various members of the staff to the tune of Gershwin's "It Ain't Necessarily So" (beginning in verse 1 with "Little Avery is small, but oh my!"). This collection of amateur poetry has been preserved because one of Avery's longtime associates, Ernest Stillman, undertook after the dinner to have it printed at a press he owned. The pages were bound together with a front sheet bearing the title that was given to this set of verses:

FORTUNE FAVORS THE PREPARED MIND

or

YOU'VE GOT TO HAVE IT IN YOUR GENES

Moral: Go change your genes!

As work on the enzyme progressed during the fall, I decided that I had aquired enough information to present it at one of the Friday afternoon staff meetings. My talk was on December 15, 1944, just a year after Avery's presentation before the same group, and I gave it the title: "Isolation and Purification of Desoxyribonuclease and Its Action on the Transforming Substance of Pneumococcus." I felt that I had to write out my speech in detail, because I didn't trust myself to keep the presentation clear and concise if I were to talk with only an outline, and I still have the original of the triple-spaced typescript that I used on this occasion. It was a rather dry and technical talk, describing the purification and properties of the enzyme and then emphasizing its effects on the pneumococcal transforming DNA. I ended this part of the presentation with a fairly explicit statement on what I thought the results implied: "Although ultimate purification of the enzyme has not as yet been achieved by the use of crystallization methods, the purity and activity of the present preparations are sufficient to provide the conclusive confirmatory evidence that was sought concerning the desoxyribonucleic acid nature of the transforming substance."

I then tacked on a discussion of an aspect of our research on the transforming substance that I have not previously mentioned in these pages. It was omitted primarily because it had no direct bearing on the main body of the work directed toward determining the chemical nature of T.P., but I have the additional excuse of having lost all of the laboratory notes on the numerous experiments that we had carried out on this phase of the problem. I had adopted MacLeod's procedure of segregating notes in separate manila folders on the basis of subject matter, and this set got misplaced after I had used the material while writing a paper on the subject, which I did early in 1945. These notes were then irretrievably lost, and I thus have no reminder of the details and timing of these studies, although I do remember that the first observations were made during my first year with Avery and that the work was carried out intermittently for the next few years. This line of research began with the observation that the addition of ascorbic acid—vitamin C—to our pneumococcal extracts resulted in complete loss of transforming activity. We explored this inactivating effect further in the hope that it might be of help in identifying the transforming substance or in determining the chemical basis for its specificity. The work never paid off in this respect.

We found that the inactivating effect of ascorbic acid was dependent on its undergoing oxidation with the formation of peroxides that served as the active agents, and a number of organic compounds that undergo the same kind of autoxidation were shown to be equally effective in inactivating T.P. This action could be blocked by excluding oxygen from the system or by incorporating a reducing agent, such as the amino acid cysteine. A more tantalizing finding was that transforming DNA which had been rendered totally inactive by treatment with ascorbic acid could be restored to almost full biological activity by subsequent exposure to reducing agents. It was apparent, therefore, that the oxidative inactivation was reversible if it had not progressed too far, and it was this aspect that I stressed in the staff meeting talk—contrasting the irreversibility of inactivation by DNase with the reversible inactivation by ascorbic acid. My paper describing the latter aspect, published in the *Journal of Experimental Medicine* in May 1945, was entitled "Reversible Inactivation of the Substance Inducing Transformation of Pneumococcal Types."[3] I was unable to claim that these observations provided any support for the thesis that the transforming substance was DNA.

My presentation of the DNase story at the staff meeting before the work was completed and the paper written had reversed the procedure we had followed a year earlier when our paper had already been submitted for publication before Avery gave his talk. I had decided that I would like to publish my findings on the purification and properties of the enzyme in the *Journal of General Physiology* (another Rockefeller Institute journal) where all of the work of Northrop and Kunitz had appeared, and I felt that this would require some further strengthening of the quantitative and biochemical data. In addition, I had not given up hope of being able to crystallize the enzyme before writing up my results. Accordingly, much of my effort in the laboratory through the summer of 1945 was devoted to making more preparations of DNase so that I could bring the study to some reasonable stopping point. I had also prepared a rabbit antibody to the purified DNase and needed to do more experiments to show that it reacted specifically with the enzyme and also that it would inhibit its action in breaking down DNA.

When I finally got around to writing the paper in the fall, I had an experience of a kind that must sooner or later happen to nearly everyone engaged in scientific research—I discovered that I had been scooped. In a review article on enzymes I found a reference to a German paper, published in 1941 in the *Zeitschrift für physiologisches Chemie,* that seemed to deal with the purification of DNase.[4] Due to the interruption of communications by the war, this paper was not available in our library or other libraries in the New York area, but I found on inquiry that I could obtain a photostatic copy from the office of the Alien Property Custodian in Washington. I quickly did this and discovered that the German workers, although they called their enzyme by a different name (thymo-polynucleotidase), had gone over much the same ground that I had. They had also apparently begun with dried pancreatic extracts and then turned to the use of fresh pancreas and acid extracts by the Northop-Kunitz procedure. Our two studies had much in common, including failure to crystallize the enzyme, although the German workers touched upon some points that I had not considered and conversely my work dealt with some matters not included in their paper. Their method of measuring enzyme activity was so different from mine that I was unable to make any direct comparison of the relative activity of our final products. In the end I had to write an addendum to the manuscript, describing the results of the German study and noting that in the light of their work part of my paper represented independent confirmation of their results.[5]

This example of the impact of a breach in scientific communication no doubt ranks among the more trivial of the destructive consequences of World War II, but it caused us some loss of time and a great deal of extra work. Had we had access to the German paper at the time it first appeared, it would have been a simple matter to have prepared some purified DNase and tested its action on transforming DNA before writing the first paper in 1943. There can be little doubt that this would have bolstered our evidence that transforming activity resided in DNA. As matters now stood, Fess and I decided to report our results on the action of our DNase on the transforming substance in a short communication in the *Journal of Experimental Medicine,* and it was ready for submission on October 10, 1945, just a few days after my paper on the enzyme had been sent to the *Journal of General Physiology.* It was given the same general title as our 1944 paper ("Studies on the Chemical Nature of the Substance Inducing Transformation of Pneumococcal Types") and designated as paper II, with the subtitle: "Effect of Desoxyribonuclease on the Biological Activity of the Transforming Substance."[6] It gave the details on the kind of experiments I have described on the inactivation of the transforming substance by the enzyme, including the differential effect of citrate on the Mg^{2+}- and Mn^{2+}-activated enzyme and the fact that a half-hour exposure to the purified preparation at concentrations of less than 0.01 microgram per cubic centimeter caused total destruction of the T.P.

In the discussion we tried to emphasize our view that the enzymatic evidence "leaves little doubt that the ability of a pneumococcal extract to induce transformation depends upon the presence of a highly polymerized and specific form of desoxyribonucleic acid." We then added a paragraph directed at the kind of criticism that we had been told was being voiced by Mirsky (he had said nothing to us directly; in fact, there was no longer any communication):

> The objection can be raised that the nucleic acid may merely serve as a "carrier" for some hypothetical substance, presumably protein, which possesses the specific transforming activity. Depolymerization of the nucleic acid would according to this hypothesis, destroy the effectiveness of the essential carrier and thus result in a loss of biological activity. There is no evidence in favor of such a hypothesis, and it is supported chiefly by the traditional view that nucleic acids are devoid of biological specificity. On the contrary, there are indications that even minor disruptions on the long-chain nucleic acid molecule have a profound effect on biological activity. Thus, treatment of the transforming substance with concentrations of desoxyribonuclease so small that only a slight fall in viscosity occurs causes a marked loss of biological activity. It is suggested that the initial stages of depolymerization which are reflected by only minimal changes in the physical properties of the nucleate are sufficient to bring about destruction of specific activity.

In essence, we felt that the burden of proof had been shifted to those who suggested that the transforming substance was *not* DNA. We agreed that the enzymatic studies did not throw any light on the chemical basis for the specificity of the nucleic acids but merely confirmed that such a basis must exist. We recognized the importance of this by writing: "It remains one of the challenging problems for future research to determine what sort of configurational or structural differences can be demonstrated between desoxyribonucleates of separate specificities."

In this flurry of publication activity there was yet another manuscript, submitted at the same time and designated as paper III in the series on T.P. The subject dealt with another dividend growing out of the work on DNase and represented our other major area of laboratory activity during the preceding months. It had occurred to me that since citrate was such a powerful inhibitor of pancreatic DNase it might conceivably have a similar effect on the pneumococcal enzyme that had long created a problem because of its ability to destroy the transforming substance. I made a crude pneumococcal enzyme preparation and determined that its action on DNA—like that of the Mg^{2+}-activated pancreatic enzyme—was totally blocked by citrate. This immediately suggested that in citrate we had in hand the kind of inhibitor that MacLeod had sought when he was trying fluoride at the time he and Avery had resumed full-time studies on transformation in the fall of 1940. It seemed to me that it ought to be possible to increase the yield of pneumococcal DNA by returning to the old Alloway procedure of lysing the living organisms with deoxycholate as long as sufficient citrate was present in the suspension of organisms at the time of lysis. This worked like a charm. Within less than a half-hour after adding a dash of deoxycholate to a suspension of type III pneumococci in citrate, the organisms had all dissolved to yield a highly viscous, translucent solution that could immediately be subjected to the first stages of purification by shaking with chloroform by the Sevag procedure. None of the old problems of loss of biological activity was encountered, and the yield of transforming DNA was several times as large as that obtained after extraction of heat-killed cells.

It was necessary to modify the purification process a bit, however, since the extract prepared by dissolving the organisms was very different in character and contained practically everything that had been present in the living cells. The amount of RNA, for example, was greatly in excess of that in our extracts of heat-killed cells, and I coped with this by using RNase at the same time that I treated the material with SIII enzyme, dialyzing the mixture during the digestion process so as to get rid of the enzymatic split products. This seemed to work, and after carrying out the fractionation steps that depended on the fibrous nature of the alcohol precipitates of DNA, we had a product with activity comparable to that of our best preparations obtained by the old method. The surprise came on analysis of the material, when we found that it was heavily contaminated with the somatic C polysaccharide, a major constituent of the cell wall of the pneumococcus. This had never been a problem before, presumably because the amount of this polysaccharide released from the heat-killed cells was very small and in a different form from that found after lysis of the entire organism.

The removal of the C polysaccharide from the transforming DNA depended on rediscovery of the merits of calcium-alcohol precipitation, with which Colin MacLeod had experimented four years earlier when he had come close to hitting upon a powerful method for separating out the transforming substance. In the presence of excess calcium ion, the addition of a very small amount of alcohol is sufficient to bring DNA out of solution in the form of a tangled mass of fibers. This precipitation is complete when an amount of alcohol equivalent to one-fifth the volume of the solution containing the DNA is added (for a final alcohol concentration of about 16 percent, or less than that of a fortified wine), and the fraction thus obtained has all of the transforming activity. However, the C polysaccharide remains in solution under these conditions, making its removal from the DNA a relatively simple process. In this way, we could get from a single 50-liter batch of pneumococci up to 80 milligrams of purified DNA (as opposed to 45 milligrams from 200 liters of heat-killed cells), which was comparable on analysis to our products prepared by the lengthy and less efficient method we had published previously. We had already known that the extraction of DNA from heat-killed cells was far from complete. I had saved the residual cells from most of our preparations and reextracted them later at a somewhat higher temperature, showing that one could still get appreciable amounts of active transforming DNA from this worked-over source.

The new method also allowed us to consider seriously the extension of our findings to the isolation of transforming DNA from pneumococcal types other than type III. We did not have a reagent like the SIII enzyme for degrading the capsular polysaccharides of the other pneumococcal types, and so we had to be selective in picking suitable organisms. Types I and

XIV, for example, proved to be like type III in having acidic capsular polysaccharides that were not separable from DNA even by the calcium-alcohol method. On the other hand, we found that the polysaccharides of types II and VI behaved in the same way as the C polysaccharide so that purified DNA could be prepared from these types. These DNAs were active in the transforming system, inducing our R strain to produce type II and type VI capsules, respectively, and thus establishing on a more general basis that the pneumococcal transforming substance is DNA. The experiments with type II were not, strictly speaking, an example of transformation, since the R strain, R36A, had originally been derived from a type II organism. However, the implications are the same, because as noted earlier R36A had totally lost its capacity to revert and could be induced to make type II capsules again only through specific transformation.

The additional studies were brought together in paper III, which bore the subtitle: "An Improved Method for the Isolation of the Transforming Substance and Its Application to Pneumococcus Types II, III, and VI."[7] To my way of thinking, the findings described added further support to the thesis that it was the DNA itself that carried the transforming activity. Armed with information about the properties of the enzyme that degrades DNA, we had been able to return to the problem of loss of activity during extraction of the transforming substance—a problem that had plagued workers in the field since the initial observations of Griffith—and show that it was readily solved by adding an inhibitor of DNase. When one considered the combined data of our original paper, the direct effect of purified DNase, and now the demonstration that DNase inhibitor would protect the T.P. during lysis of the organisms, there was not much room left for the skeptics to advance sensible alternatives to the view that DNA was the active substance in transformation. There is some evidence to suggest, however, that papers II and III (which finally appeared in February 1946, just two years after paper I) were not very widely read. They were infrequently referred to in discussions of the transformation work, and I found that it was commonly assumed that data on the action of a purified DNase on the transforming substance had been included in the first paper. I will explore more fully the general reaction to our results in the last chapter.

A final note about the crystallization of DNase. My efforts had come to naught and just before I left the Avery laboratory in 1946 I approached Kunitz with the suggestion that he take up the problem. I got the impression that he was eager to do this, having held off up to this time because he considered it my property. I took some comfort from the fact that he didn't find it easy either, but in the end he succeeded and was able to publish a preliminary report on the crystallization of deoxyribonuclease in *Science* in 1948.[8] Hotchkiss, who had come to Avery's lab as I left and continued to work on transformation, showed that the crystalline enzyme was more potent than my partially purified DNase in destroying transforming activity. When Kunitz published two full-length papers on crystalline DNase in the *Journal of General Physiology* in 1950,[9] he was very generous about my role in this business when he wrote: "The present studies, which led to the isolation of the enzyme in crystalline form, should be considered as a continuation of McCarty's work, since Dr. McCarty not only suggested to the writer that he enter the field but also cooperated in the initial stages of these studies." I had visited him on a couple of occasions while the job was in progress and, in addition to the vicarious enjoyment of seeing my baby come out in the form of crystals, I had an opportunity to get some idea of how he did it. Patience was hardly the word to characterize his approach to the problem. Meticulously, and with a highly systematized procedure, he tried the effect of a whole series of variables (salt concentration, acidity, temperature, etc.), each of which was changed independently in small increments. It was laborious, but his sweeping approach just about assured him of ultimate success.

My consolation prize for having failed at crystallization of DNase was to come up unexpectedly with crystals of another substance of interest in the laboratory, C-reactive protein. We always checked our chest fluids for their possible content of C-reactive protein, and early in 1946 we received a large volume of fluid (from one of the navy patients with pneumonia in our hospital) that proved to have a high concentration of the substance. Since the fluid proved to be useless in the transforming system, I decided to isolate the protein for possible future studies. It was in the course of this process that I obtained it in crystalline form, more or less accidentally, although I am sure that my experience with the pancreatic enzymes had sensitized me so that I knew what to look for. Having a protein from human serum in the form of beautiful rhomboid plates[10] was an exhilarating event in itself, but it also led to a renewed interest in this substance that fostered future studies.

XI

THE FINAL MONTHS

As THE WORK ON DNase and its application to the problem of transformation neared an end, Fess and I had had a number of discussions about what line of research we would tackle next. There were a number of unanswered questions concerning what actually went on in the transforming system when one added transforming DNA to the growing culture of rough pneumococci, and there remained the challenge of whether there was anything useful we could do to dispel any of the lingering doubts that DNA was itself the transforming substance. Among the oldest of the unsolved problems was that of the serum factor and why it was necessary to add it to the transforming system. We still had no clue as to the purpose it served. It seemed likely, however, that it was involved in providing the right conditions for the interaction of the rough pneumococci with the transforming DNA and that we might learn more about this interaction if we could discover the nature of the serum factor. I was thus led into some unproductive research, referred to earlier, in which I repeated and extended the kind of experiments on the fractionation of the serum factor that Colin MacLeod had been involved in some ten years before. The reasons for the failure of this approach are probably not worth discussing in detail.

The elusive serum component was later shown by Hotchkiss to be the single most abundant protein in serum—albumin—which appeared to act by protecting the pneumococci from certain toxic substances in the growth medium.

My frustration during this period was more than matched by Fess's depressed state of mind, which created an uneasy atmosphere in the laboratory. He had always been susceptible to swings in mood, but the periods of apparent discouragement and depression had in my experience been of short duration. I was told by Horsfall, however, that prior to his thyroidectomy in 1934 Avery's natural ebullience had been quite understandably suppressed for a long period. I believe that his difficulties in 1945 were much more complex in origin. In the first place, after the long and dedicated search for the nature of the transforming substance, the completion and publication of the DNA paper followed by the confirmation with DNase gave an aura of anticlimax to our current activities. Furthermore, he continued to be plagued by nagging doubts about whether we were right, doubts that were clearly not ameliorated by the reservations expressed by others, such as Mirsky, and the apparently rather restrained acceptance of the thesis advanced in the 1944 paper. I suspect that he was also suffering some mental conflict in relation to his commitment to retire and move to Nashville to join his brother, Roy. The reasons that he had advanced for not making the move in 1943 were no longer valid, but we were all convinced that he had little enthusiasm for leaving Rockefeller and New York. He was obsessed with the idea that he should leave, however, and frequently reiterated his observation that "A man should know when he's through, move on, and get out of the way of the young."

On top of all this, much of the work going on in the laboratory was not the kind that he felt comfortable about participating in personally. As a result, he had a great deal of free time that he had difficulty in filling with other activities. He frequently went off to visit some of his former associates—

René Dubos, Frank Horsfall, or Rebecca Lancefield—in their laboratories where he would be brought up-to-date on the recent developments in their research. There was a limit to the amount of time he could spend in this fashion with busy people who were intent on their own activities and inevitably there were many hours when he was at loose ends. He would come and sit in the chair beside my desk, saying very little but effectively holding me immobile to share his gloomy outlook. If I made a move to get up to do something in the lab, he would make a gesture with his hand, raising one finger as though he were about to say something, a strategem that worked very well to keep me at the desk even after I had learned by experience that he was not likely to follow it up with conversation.

As time went on I found it difficult to cope in this situation with the necessary restraint and good humor, and I'm afraid that I was not nearly as patient with Fess as I should have been. It was particularly aggravating to see how quickly he could turn off the gloom if the social circumstances required it. On occasion, while sitting at my desk, he would receive a telephone call from a friend like Shosho MacLeod, Colin's wife, and immediately undergo a remarkable alteration. His face would light up, become animated and smiling, and his voice would acquire a new timbre, expressing interest and enthusiasm—features of his personality that I had not seen or heard for days. The reversal was just as sudden and when he had finished the conversation and hung up the phone, his face fell abruptly with the return of his apathetic expression, just as if it had been triggered by breaking the connection. He never gave any direct indication that he was aware of my irritation, which must have been all too obvious at times, but he did make a couple of peace offerings which suggested that he had some insight into the problem and the stress that his partial paralysis imposed on me. In April 1945 he presented me with a copy of the recently published second edition of Karl Landsteiner's famous book *The Specificity of Serological*

Reactions. Then, about two months later on my birthday, he gave me a copy of René Dubos's first book, *The Bacterial Cell*, which had just appeared. He had never done anything like this before, and I interpreted the gifts as tokens of apology, although the brief inscriptions in the books—"To Mac from Fess"—provided no suggestion of this.

Our relationship gradually improved without ever quite reverting to the easy camaraderie that prevailed during the heat of the search for the nature of the transforming substance. An event that almost certainly fostered the improvement was the arrival of a new face in the laboratory early in the summer of 1945. Harriett E. Taylor had just received her Ph.D. in genetics from Columbia University and had been awarded a postdoctoral fellowship from the National Research Council to come for training in the laboratory. A bright, talented, and personable young lady, Harriett was an asset to the laboratory from the very beginning. In retrospect, I believe that her joining our studies at this juncture also brought a boost in morale through the implications of her having chosen the lab for her first postdoctoral experience. Here was at least one young biologist, trained in classical genetics and steeped in its traditions, who clearly saw the significance of the studies on pneumococcal transformation and accepted the evidence that DNA was the carrier of genetic information. Indeed, she looked upon transformation as the wave of the future in genetic research.

Harriett was soon taking an active part in all of the experimental work, including the unproductive attempts to identify the serum factor by fractionating and purifying it, and at the same time she initiated some studies of her own on more genetic aspects of pneumococcal transformation. She was an excellent worker and adapted quickly to her new environment, which must have been quite different from anything she had known in the biology department at Columbia. Even the frustrations of the serum factor problem did not serve to dispel the new sense of harmony that had settled on the lab.

Fortunately, some of our efforts were more successful and rewarding. It had occurred to me that the purified DNase might provide a means of finding out more about what was going on during the process of transformation. Since one could easily add the enzyme to the transforming system in amounts that would destroy any free DNA in a matter of seconds, it seemed possible to determine how long it took the rough pneumococci to take up the DNA after they had been inoculated into the system. Once the DNA had been taken up by the pneumococci and the process of transformation initiated, the addition of DNase could no longer have any deleterious effect. To test this idea, I started out with a large group of tubes of the transforming medium, each of which contained a liberal dose of type III DNA and the usual inoculum of R36A. On overnight incubation, transformation would have occurred in 100 percent of these tubes, if they were not tampered with. My tampering consisted of temporarily removing sets of four tubes from the incubator at half-hour intervals and lacing them with an amount of DNase that was capable of wiping out the viscosity of a DNA solution in less than a minute. The incubation of the whole collection of tubes, which had been subjected to DNase treatment at different times, was then continued overnight for the completion of the transformation reaction.

The results of this kind of experiment were clear cut and consistent. If DNase were added to the system at any time during the first 3 to 3.5 hours, transformation was completely blocked and only the R organisms from the original inoculum were found on subculture. On the other hand, if the addition of DNase were delayed until four hours or later, it had no effect on the reaction and transformed type III cells were found in all tubes in the usual numbers. This seemed to be telling us that, in the transforming system we had been using for the past several years, the presence of the T.P. was irrelevant for the first few hours of growth. The transforming DNA simply sat there passively, waiting for the rough pneumococci to get in the mood to receive it. We could check this by growing the R organisms for 4 hours in the transforming medium without DNA and then determining how quickly they could take it up after it was added to the culture. Again using DNase to eliminate unreacted DNA, but at shorter time intervals, we found that the R organisms in these 4-hour cultures were able to complete the transaction and proceed with transformation after exposure to T.P. for as short a time as 5 minutes. Obviously, some change had taken place in the pneumococci during the first few hours of growth that permitted them to interact rapidly with DNA in their environment.

This alteration in the R organism that seemed to be required for transformation was transient in the sense that, if incubation of the culture in the absence of DNA were continued beyond 4 hours, they gradually lost the ability to take it up rapidly when it was added. On the other hand, Harriett Taylor found that if a 4-hour culture were quickly chilled and then refrigerated overnight, the pneumococci retained the property of rapid interaction with DNA. Whatever the modification of the organisms was, therefore, it could be preserved for a considerable period of time if further growth and metabolism of the living cells were temporarily suspended by reducing the temperature. We had little information as to what was going on to bring about this alteration in the behavior of the organisms, leading to a condition that came to be known as "competence," but the work of others a decade or two later threw some light on the matter and showed it to be a complex biological phenomenon.

The one thing that we knew was going on, of course, was that during the 4 hours of initial growth the population of pneumococci was progressively increasing, from the few thousand organisms of the inoculum to about one or two million. It was easy to show, however, that population size had little or nothing to do with competence. As a matter of fact, in the course of establishing this we found that with our improved transforming system and purified transforming substance the success of transformation had none of the marked dependence on inoculum size that Dawson and Sia had described when they first succeeded in getting transformation in the test tube, an advance that was achieved only after a thousandfold reduction in the number of organisms inoculated. In our system, inocula of millions of cells were compatible with transformation, but even with these large numbers a few hours' growth in the transforming medium was required before they became competent.

Although these experiments on the conditions required for the uptake by the rough pneumococci were not undertaken with the idea that they might have a bearing on the part played by serum factor in the transforming system, it soon became evident that this was the case. The organisms became competent and capable of rapid DNA uptake only when the 4-hour growth took place in the complete medium, which contained chest fluid to provide the serum factor and anti-R antibody. Growth in blood broth or in medium containing only anti-R, in which the organisms multiplied perfectly well, would not do the trick. This implied that, whatever the nature of the serum factor might be, one of its principal functions was to provide the appropriate conditions that would allow the pneumococci to undergo the changes necessary for the assimilation of the transforming DNA. This served to sharpen the focus of the studies on the troublesome serum component and changed the orientation of subsequent experiments.

During this period we also looked at other aspects of the transformation reaction. For example, we were interested in getting information on the frequency of transformation—that is, how many of the pneumococci exposed to transforming DNA actually went on to produce type III capsular polysaccharide? We looked at this by diluting out competent organisms after they had been treated briefly with DNA and determining the minimal number that would yield type III colonies after transfer and outgrowth in complete medium. Our best results indicated that from 0.5 to 1 percent of the organisms in a 4-hour culture were able to carry the reaction to completion. While this was probably a minimal figure, since the manipulations tended to decrease the efficiency of transformation, the data gave us some further insight into what was going on in the tubes of our test system.

I had gradually become involved in giving more and more talks about our work, with the result that the laboratory experiments were no longer the strictly full-time endeavor that they had been. One of the motives for these talks was to make our findings better known among groups that might not ordinarily read the *Journal of Experimental Medicine*. Without having any record of the exact date, I can remember making a trip to Cold Spring Harbor, New York, to give a presentation to the members of what was then the department of genetics of the Carnegie Institution. I also talked about the DNase before the Enzyme Club of New York City, which in those days met regularly at the Faculty Club of Columbia University. These reports on our work were good training for me, but I have little idea how effective they may have been in spreading the word about DNA.

Contrary to my impression at the time, I discovered long afterward that Avery had taken part on at least one occasion in these missionary presentations. He was so secretive about it that I had no inkling of his talk until I saw it referred to in a paper by Wyatt in 1975.[1] It was given on the occasion of the meeting of the American Association for the Advancement of Science held in Cleveland in September 1944. Avery was vice-president of the Section on Medical Sciences of the Association and presented a paper that, according to a report of the meeting published later in *Science*,[2] was entitled "Experimental Induction of Specific and Heritable Changes in Pneumococcal Cells." True to his pattern of behavior with talks of this kind, Avery did not allow the paper to be published. The brief summary, by the secretary of the section, that was included in the report of the meeting made no mention of nucleic acid and revealed a rather restricted view of the work in its final sentence: "The far-reaching implications of these findings in the field of microbic life were considered in detail." I had not seen this report in *Science* at the time, and in checking the reference recently I noted that I had also missed an item two weeks earlier in the same periodical reporting on the award of the Gold Medal of the New York Academy of Medicine to Avery on October 5 of that year.[3] In his citation the president of the Academy had come somewhat closer to the mark than the A.A.A.S. section secretary: "You have . . . isolated the 'transforming principle' as a thymonucleic acid. This discovery has very far-reaching implications for the general science of biology."

My first participation in a more formal conference came about because of Fess's proclivity for not answering letters. It is not true, as sometimes stated, that he never answered letters, but he did tend to procrastinate in replying to those that wanted something from him or requested detailed information. He had an old-fashioned, roll-top desk in his small office on the floor above the laboratory, and when it became too full of unanswered correspondence that he no longer wanted to face he would simply cover it all up by leaving the roll-top down. On more than one occasion I saw him standing at the closed desk and working on the top rather than open it and remind himself of his derelictions.

Sometime in the late spring of 1945, Fess received a letter inviting him to participate in a conference on "Intracellular Enzymes in Normal and Malignant Tissues." The conference was to be held in October under the sponsorship of the International Cancer Research Foundation and the Jane Coffin Childs Memorial Fund for Medical Research. He had long since given up participating in activities of this kind, and consequently the letter was relegated to the stack of unanswered mail and remained there when he went off to Maine for vacation. The organizer of the conference, forewarned about Avery's reputation as a correspondent, was resourceful enough to write a second letter in August in which he indicated that he had interpreted Avery's silence as meaning consent, and accordingly he was providing additional information and instructions with regard to the conference. This served to bring quick action from Fess, whose immediate response was to telephone me from Deer Isle and ask whether I would be willing to take his place at the meeting. Since I did not yet have any of his inhibitions and was still eager to spread the word about DNA, I was only too happy to accept. Very soon thereafter I received my official invitation to attend.

As indicated by the title, the conference was oriented principally toward cancer research, but one of the sessions was on "Intracellular Enzymes of Nucleic Acid Metabolism" for which I was asked to contribute a paper on the biological role of the nucleic acids. This suited me fine, since I would be able to summarize the past views about the function of nucleic acids and then emphasize the implications of our findings with the pneumococcal transforming substance. The conference was an interesting one, held in a rambling resort-type hotel in Hershey, Pennsylvania, and attended by many of the outstanding biochemists and cancer researchers of the day, seven of whom were subsequently awarded the Nobel Prize. It was a successful experience for me in that I learned a great deal during the four days of the meeting, although again I was not sure that I had managed to convince any skeptics about the genetic role of DNA. The copy of my prepared remarks shows that I tried to relate our work to the main topic of the meeting by saying: "the fact that desoxyribonucleic acid seems to play a specific role in inducing and maintaining a heritable modification in pneumococci gives rise to speculation concerning the role of nucleic acid in the transformation of a normal tissue cell into a malignant one. In some types of neoplasm, at least, the basic change may be a modification in the chemical configuration and specific action of certain nucleic acid molecules with a resultant change in the metabolism and growth properties of the cell." This bit of gratuitous speculation didn't exactly create a sensation either.

Some months later I had an opportunity to present our findings in a quite different context, this time before a group interested in viral research. Wendell Stanley had invited Avery to participate in a symposium on "Biochemical and Biophysical Studies on Viruses" to be held in April 1946. Avery of course declined and again offered me as a substitute. It was natural for Stanley to include a paper on transformation in a meeting that dealt with viruses—he had considered the transformation phenomenon to be "virus-like" since the mid-1930s, and we had referred to his published statement of this view in the discussion section of our 1944 paper. His symposium was a one-day affair held as part of the annual meeting of the American Chemical Society in Atlantic City, New Jersey. Because the other papers on the program that day all dealt with animal, plant, or bacterial viruses, I discussed the relationship of the transforming substance to viruses after summarizing the evidence for its identification as DNA. I was inclined to emphasize the differences rather than the similarities.

The ease with which the transforming substance can be inactivated by enzymatic action stands in direct contrast to most of the accumulated experience with animal and plant viruses and bacteriophage, which have been shown to be highly resistant to inactivation by nucleases, and in most instances by other enzymes as well. It may be that this is an indication of a fundamental difference between the transforming substance and the viruses, a difference that had already been suggested by the apparent absence of protein and serological activity.

I then concluded my talk with the following statement:

It will be observed from the foregoing discussion that while the pneumococcal transforming substance is virus-like in certain of its properties, there is some evidence inconsistent with its classification with the viruses, despite the diversity of this group of agents. However, if one accepts the validity of the view that the biological specificity of the transforming substance is the property of a desoxyribonucleic acid, the results of the present study serve to focus attention on the nucleic acid component of the virus nucleoproteins.

In addition to its probable role in the self-reproduction of the virus molecule, the nucleic acid may carry a specificity which is a determining factor in the ultimate structure of the virus.

It was clear that Stanley had not drawn this conclusion from our work himself, and he was not receptive to my remarks on this occasion. In a historical piece on our work, written in 1970, he said:

It is obvious that despite my 1938 writings, I was not impressed with the significance of the 1944 discovery by Avery, MacLeod, and McCarty or I would have prepared high molecular weight tobacco mosaic virus-RNA once again and tested it for virus activity despite the fact that RNA was not suspected to have genetic properties. It remained for Fraenkel-Conrat to do this important experiment in my laboratory 14 years later.[4]

By the time he made these remarks in 1970, the free nucleic acids of several viruses had been shown to be infectious and capable of leading to the production of complete virus by the infected cells. This is essentially what I was trying to predict in a vague way in the final sentence of my symposium paper.

The proceedings of neither of these meetings—the Hershey conference on cancer or the Atlantic City symposium on viruses—were published, and the message of my summaries of our work was not disseminated beyond the groups that attended the sessions. My first opportunity to reach a larger audience came about in a quite different and gratifying manner. Some of my colleagues, and I believe that Avery and Horsfall were the prime movers, had nominated me for the Eli Lilly Award in Bacteriology and Immunology. This was one of the very few research awards in the field, and only young investigators under the age of 35 were eligible; since I was in my 35th year, my nominators had no time to lose. In submitting the necessary documentation they had to include copies of the manuscripts of the DNase work, because these papers were still in press at the time the nomination was initiated. They managed to keep the whole business a secret from me so that the letter announcing my selection for the award came as a complete surprise.

The Lilly Award was administered by three national scientific societies in the field and was presented at the annual meeting of the largest of these, the Society of American Bacteriologists. The Society met that year in late May in Detroit, Michigan, at the old Book-Cadillac Hotel. Although the scientific sessions lasted for a full week, I didn't have to appear until Thursday, May 23, the day of the annual dinner at which the award was to be presented by the president of the Society. I arrived on that day, still in navy uniform, only to find that there was no room for me at the hotel. It seems that the name of the award winner was a closely guarded secret, meant to be a complete surprise until it was announced at the dinner, and the secretary of the Society, in a misguided effort to assure confidentiality, had hit upon the device of making my reservation in the name of Dr. L. A. Ward (L. for Lilly, of course) instead of mine. Unfortunately, when the supply of rooms had been exhausted earlier that week, some officious person had volunteered that she knew Dr. Ward and was sure that he was not coming to the meeting, with the result that my room was reassigned. After a short period of confusion, the secretary solved the problem by sharing his official suite with me.

I made the required address on the award-winning work before a plenary session on the morning after the dinner; that is, on May 24, about two weeks before my 35th birthday. This talk carried the title: "Chemical Nature and Biological Specificity of the Substance Inducing Transformation of Pneumococcal Types." I was allotted enough time to give a reasonably detailed review of all of the research that we had completed up to that time and to indulge in a little discussion of the implications of the findings. I did not, however, break away from our earlier reserved approach by explicitly claiming that our results established DNA to be the substance responsible for carrying genetic information in all living organisms. Rather, my discussion tended to follow the pattern of our papers and the recent symposium talks. For example, part of the comments on the implications were phrased in the following terms:

Indeed, there are certain striking analogies between the biological properties of the transforming substance and those of viruses and genes. For example, as in the case of viruses, the transforming agent acts only on susceptible living cells; it is transmissible in series and can subsequently be recovered in amounts far in excess of that originally used as inoculum. As in the case of genes, the transforming substance behaves as a heritable unit in that it produces predictable and durable alterations in cellular structure and function and is reduplicated in daughter cells through successive generations. It intervenes in the metabolism of the R cell, giving rise to the synthesis of a new capsular substance, which in turn endows the cells with distinctive and biologically specific characters not possessed by the parent strain. Although the validity of these analogies may be questioned, they serve to underline the possible implications of the phenomenon of transformation in the field of genetics and in virus and cancer research.

On the other hand, I tried to be fairly forthright about our conclusion that the transforming substance was indeed DNA:

Certainly there can be little doubt that desoxyribonucleic acid must be present in its intact, highly polymerized form [for transformation to occur], and when all of the evidence is considered it appears extremely unlikely that small traces of some other specific substance, such as a protein, could be responsible for the manifestation of transforming activity.

This review was published by the Society in one of its journals, *Bacteriological Reviews*,[5] with unusual speed. It appeared that same summer, thus providing some additional exposure of our results to a bacteriological audience. The Lilly Award also provided me with a handsome silver medal and a check for $1000. The medal I still have as a memento of this exhilarating time, and the check was promptly used to buy my wife a piano, a luxury that we had been unable to afford during the early years of our marriage.

Another source of distraction arose in the late winter of

1946 as all of us in the naval reserve unit were anticipating the termination of our navy service and return to civilian life. Most of us did not have regular appointments at the Rockefeller Institute and were uncertain as to what we would be doing when it reverted to peacetime status. Earlier in the war, Dr. Rivers had been asked to organize an advance naval research unit for the South Pacific, made up of a few members of the Rockefeller group and a number of other investigators recruited for their specific area of expertise. He assumed command of this unit and moved out to Guam when it was established there, leaving Horsfall in charge of the group that remained at the Rockefeller Hospital. He spent all of 1945 at this outpost, and his extensive correspondence during this period with Horsfall and the director of the Institute, Herbert Gasser, indicated his concerns about the future of the hospital and its need for reorganization after the war. On his return in January 1946, these matters claimed his top priority.

One of the most urgent of Rivers's problems was the imminent retirement of the remaining senior members of the hospital staff. The first of these was Homer F. Swift, head of a laboratory concerned with streptococcal infections and rheumatic fever, who was scheduled to retire in June 1946. Because of the organization of the Institute there was no requirement for retaining a laboratory in a specific area of investigation on the retirement of the chief; but in the case of rheumatic fever, Rivers felt that the disease continued to be of sufficient importance to make it desirable to do so. While he must have considered a number of possible candidates for a successor to Swift, I have never had any information on this point. I was taken completely by surprise when he called me to his office in March and asked if I would be interested in taking over the Swift laboratory on July 1.

The proposal did not include promotion to the rank of tenured Member of the Rockefeller Institute, but it was clearly implied that this would be forthcoming if I proved myself in this new area of research. I had no qualms about the prospect of probationary status, since the offer seemed to me to represent a real chance to get a permanent place at Rockefeller, a goal that had seemed so unattainable that I could hardly admit to myself how much I wanted it. A more important question was whether the problem of streptococcal infections and their relationship to rheumatic fever represented an area in which I was interested in devoting my future research efforts. I had few, if any, reservations about this. I had had extensive experience with the care of young patients with rheumatic fever during my pediatric residency at Johns Hopkins and had considered the disease to be among the most challenging problems in the realm of infectious diseases. In addition, Swift's laboratories occupied the other half of our floor of the hospital building, so that my interest in the problem had been constantly reinforced through contact with members of his group during my years with Avery. In essence, the unsolved puzzle was how the common streptococcal sore throat managed to lead to the delayed appearance in many patients of rheumatic fever, with its frequent severe involvement of the heart. Most of the serious heart disease of young adults at that time could be traced to the occurrence of one or more attacks of rheumatic fever in childhood. It was evident that an essential line of attack on the problem was to learn more about the causative streptococci, making it a problem in medical microbiology that certainly appealed to me.

There was one difficulty. How could one even consider turning from the path of research opened up by the DNA discovery? No doubt the lure of being on my own and having a chance at a permanent Rockefeller appointment were factors, but there were other considerations. I was not trained as a geneticist and was little attracted to pursuing the studies along genetic lines. We had talked about attempts to transform characters other than capsule formation in order to extend the significance of the phenomenon, and we had even discussed possible models for demonstrating DNA transformation in higher, nonbacterial species. However, we made no move in this direction, and the former approach was left to be exploited successfully by Hotchkiss and others. (It was not until recent years that DNA transformation has been accomplished in higher organisms.) My own preference was to continue with a biochemical attack. We knew that DNA could be extracted from R pneumococci in the same amount and with the same general properties as that obtained from the S organisms, but it was without effect in our capsule transforming system. Our interpretation of this finding was that most of the DNA in our purified transforming substance was concerned with innumerable other functions and properties of the pneumococcal cell, like that from R pneumococci, and that only a small fraction was involved in directing the synthesis of capsular polysaccharide. I thought that it should be possible to prove this by fractionating the material in order to obtain purified, or at least highly enriched, preparations of the DNA molecules responsible for capsular transformation. I made only tentative and completely unsuccessful attempts to separate out the active fraction of DNA, and it is just as well that I did not become committed to a more serious effort. Even with today's knowledge and techniques it would be a monumental task.

All in all, it seemed that I was coming to a good stopping point in these studies, and I did not delay long in telling Dr. Rivers that I would accept the position. He sent me to Dr. Gasser for the purpose, I am sure, of getting his concurrence with the appointment. This interview didn't seem to be going particularly well until I volunteered the information that I was very much interested in the problem of rheumatic fever and eager to begin research in the area. His comment was, "That's what I've been waiting to hear you say," and it turned out that little more had to be added. Rivers had of course discussed the matter with Avery in advance, which paved the way for me when I went to Fess to seek his advice in the matter. Fess would never tell one what to do in these situations, but he was very good at analyzing the pros and cons, thus clarifying the basis for a decision. I got the impression that he considered the offer a great opportunity for me and that he was neither surprised nor disappointed when I accepted it.

After it was settled—sometime in April, as I remember it—that I would be moving to the other end of the laboratory floor on July 1, new plans were made for continuation of the studies on pneumococcal transformation. It was arranged that Rollin Hotchkiss would join Fess and Harriett Taylor when I departed. Hotchkiss, a skilled and experienced biochemist, had long had an intense interest in the problem of transformation and had, in fact, asked Fess to let him participate in the work in 1938. Fess had replied something in the nature of "Not now," a response that may have been motivated by the fact that this was during the period that the problem had been temporarily set aside. In view of this history, which I was not aware of until many years later, it seems likely that Hotchkiss was not any more enthusiastic about my appearance on the scene in 1941 than MacLeod was. Like MacLeod, however, he never showed evidence of this in our personal relations, and was always a helpful and cooperative colleague.

I had a few weeks to bring my experimental work on transformation to some kind of reasonable conclusion and not leave behind any untidy overlaps with the new regime. There was one more conference still ahead of us. We had been invited to participate in a Cold Spring Harbor Symposium on "Heredity and Variation in Microorganisms" to be held that June, and we had yet to prepare for it. Fess had as usual declined the invitation and offered me as substitute, and I conceived the idea of using this forum for the presentation of all of our recent research in order to avoid leaving behind pieces of the problem in which I felt I still had a significant stake. Quite simply, I thought it would be better if I could make a clean break. The paper I put together was entitled "Biochemical Studies of Environmental Factors Essential in Transformation of Pneumococcal Types," with Harriett Taylor joining Avery and me as co-author.[6] I included all that I thought was salvageable from the work on serum factor, together with the data on the time of uptake of DNA by R pneumococci and some experiments, which I have not described in these pages, that gave some indication of what the R antibody was doing in the transforming system. The paper was prefaced with a brief historical introduction in which we reiterated our earlier view that our work "had established beyond reasonable doubt that the active substance responsible for transformation is a specific nucleic acid of the desoxyribose type," and then tried to explain the rationale for our interest in the "environmental factors." It may have been a tactical error to focus on these ancillary matters rather than presenting a detailed review of the evidence for the DNA nature of the transforming substance as I had done for the Lilly Award lecture. There were a number of geneticists at the symposium, and I suspect that most of them had not heard the facts on which we based our conclusion.

In any event, I did not get the impression that many of the geneticists there wholeheartedly embraced the implications of our findings for their own genetic studies. Harriett and I both attended the meeting, and I believe that she may have been a more effective proponent of our views than I was, although I have no record of specific successes in this regard. At the end of the session at which I had presented the paper, one of the geneticists actually came up to me and said: "Now that you fellows have shown that nucleic acid is not responsible for transformation, why don't you get to work and find out what really is?" I recovered quickly enough from this sally to reply that I was under the impression that this was exactly what we had done—shown that it was DNA, thus abruptly ending the conversation. He was a hard case, however, and his subsequent writings indicated to me that he didn't have a clear understanding of the phenomenon of pneumococcal transformation, let alone the work that we had done on the transforming substance.

These Cold Spring Harbor Symposia were held annually

on selected topics of what was called "quantitative biology," and the entire proceedings of the meeting were published each year in a hard-cover volume. The delay in publication was great enough in 1946 so that by the time we had galley proof Hotchkiss already had some of his evidence that pure preparations of serum albumin would effectively replace the serum factor in the transforming system as long as a little R antibody was supplied from some other source. This suggested that what we had written about a globulin fraction (the other major class of serum proteins) playing a role in serum factor was simply wrong. He called this to the attention of Avery who in turn let me know that we had a problem. As Hotchkiss put it some years later, Avery "refused to borrow our facts [Hotchkiss's and Harriett Taylor's] to decorate their paper, but neatly protected the readers by changing every 'globulin fraction' in the manuscript to 'protein fraction.' "[7] It was all the more clear to me, however, that I had made a mistake in trying to push our unsatisfactory serum factor work into print just in order to leave no loose ends of the problem behind when I moved out of the laboratory.

That move occurred on July 1, 1946, without any significant disruption of normal activity. It was also the first day that the members of our naval reserve unit were eligible for discharge from the service, and I found time to go through this process as well. The assumption of my new responsibilities and my return to civilian life thus occurred on the same day.

XII

AFTERMATH

MANY DIFFERENT OPINIONS have been expressed about the reception in the scientific community of our reports on the biological activity of pneumococcal DNA and about the extent to which the work was unappreciated or ignored. Since we had our own views on the subject, we were naturally interested in these opinions as they appeared but never made any public comments on them. In discussing the matter here, I have no hope of resolving the numerous issues that have been raised. However, in reviewing this part of the history, I can call attention to a number of factors that certainly influenced the reception of the pneumococcal DNA story and give my own personal assessment of the degree to which its acceptance was retarded.

In the first place, there is little question that the date of the appearance of our first paper in February 1944 sharply limited the audience that it reached. Coming as it did near the height of our involvement in World War II, the paper was seen by only a fraction of the biomedical scientists who ordinarily would have been following this journal and by almost none of those who were working abroad. The later papers on the pancreatic DNase and its effect on the transforming factor did not fare significantly better, since they appeared prior to the restoration of the normal channels of scientific communication after the war. As a matter of fact, these follow-up studies did not seem to benefit from the subsequent renewed interest in the first paper, which apparently resulted from further dissemination of information about the pneumococcal studies, chiefly from the written comments appearing in review articles by biologists but also by word-of-mouth. We had originally ordered three hundred reprints of the first paper, but these were quickly depleted in the postwar period so that we had to obtain an additional three hundred copies by photo-offset to satisfy the renewed demand. On the other hand, papers II and III were rarely cited in publications dealing with the subject, even though the authors always referred to paper I. In many cases it was not clear whether the writer was even aware of the additional evidence that had come from the use of DNase.

A second factor relates to the readership of the journal in which the work was published. To be sure, all of the earlier papers from the Avery laboratory on the subject of pneumococcal transformation—that is, those of Dawson and Alloway—had appeared in the *Journal of Experimental Medicine*, but the *Journal* was not one that was read by geneticists and general biologists. Thus, those scientists who were most likely to be interested in the broader implications of the findings became aware of the work only if someone specifically called it to their attention. Under these conditions, the manner in which they heard the facts assumes considerable importance, and this has a bearing on the impact of the reservations expressed by Mirsky, as I will discuss shortly.

In addition to these immediate and specific factors that influenced full dissemination of the pneumococcal DNA story, there were elements in the general scientific climate that contributed to slow acceptance of the results. As noted in an earlier chapter, the science of bacteriology had developed almost as if it were not an integral part of biology. Those bacteriologists who dealt with variation in microorganisms had adopted their own terminology instead of that of the geneticists, speaking, for example, of bacterial dissociation rather than mutation. On the other hand, the geneticists, who had made their great strides in the understanding of genetic mechanisms through the analysis of multicellular organisms that undergo sexual reproduction, did not consider the bacteria, with their simple life cycles, presumably devoid of any element of sexual reproduction, as suitable for genetic study. The techniques of mating and crosses, for example, could not be used to establish genetic information as they were with the fruit fly or maize. New information on bacterial systems was just beginning to break down this rather artificial barrier between genetics and bacteriology in the mid-1940s, but many of the classical geneticists were still influenced by the traditional view.

Thus, the majority of geneticists were not prepared to accept information emerging from studies of the pneumococcus as having any bearing on the genetics of higher organisms, just as the majority of biochemists were still influenced by the notion that nucleic acids were monotonously alike and therefore not likely candidates for the possession of biological specificity. As pointed out previously, even though the scientific evidence on the concept of a uniform structure of all nucleic acids was neither very rigorous nor convincing, the idea had not been seriously challenged by recent chemical research and its survival was fostered by the growing emphasis on proteins as the bearers of biological specificity. The proteins were obviously tremendously diverse and it was becoming increasingly clear that they were responsible for the specificity of a vast array of biologically important molecules, such as enzymes and antibodies.

Alfred Mirsky had taken the point of view that we did not have adequate evidence to claim that the transforming substance was DNA, and he had prepared a detailed assessment of the evidence to support this contention. He did not discuss the matter with us, but we had heard by the grapevine that he was expressing his opinion frequently in personal conversations with interested individuals. Since he was widely acquainted with biologists and biochemists, this dim view of the implications of our work certainly reached many ears and undoubtedly had some influence on its reception. As far as I know, Mirsky's first public presentation at the Rockefeller Institute of his opinions on this matter was on April 12, 1946, when he gave a lecture at one of the Friday afternoon staff meetings on "Some Aspects of the Chemistry of Chromosomes." Even then he did not incorporate the topic in his lecture but summarized his views in response to a question from the floor during the discussion period. I was not present at the lecture, since this was the day that I had to be in Atlantic City to participate in Wendell Stanley's American Chemical Society symposium on viruses. Fess had sat through the whole talk without, of course, making any comment, and when he told me about it on my return the following Monday he said: "I'm glad you weren't there, Mac. You would have felt that you had to answer him." He was still happy to let the evidence speak for itself.

Because of my absence I had only secondhand information on the precise line that Mirsky followed in making his comments on the transforming substance. I knew his point of view in general, however, and I assume that what he said at the meeting was pretty much in the same vein as what he wrote later in the year when he and Pollister published their first full-length papers on their nucleoprotein studies in which he also described our single collaborative experiment. The key paragraph reads as follows:

> Avery and his colleagues have shown decisively by inactivation experiments that desoxyribose nucleic acid is an essential part of the transforming agent, and if there is actually no protein in their preparation, it would be obvious that the agent consists of nothing but nucleic acid. This is a conclusion of the greatest interest in the study of the chemical basis of biological specificity, and it should therefore be scrutinized carefully. There can be little doubt in the mind of

> anyone who has prepared nucleic acid that traces of protein probably remain in even the best preparations. With the tests now available for detecting how much protein is present in a nucleic acid preparation, it is probable that as much as 1 or 2 per cent of protein could be present in a preparation of "pure, protein-free" nucleic acid. . . . No experiment has yet been done which permits one to decide whether this much protein actually is present in the purified transforming agent and, if so, whether it is essential for its activity; in other words, it is not yet known which the transforming agent is—a nucleic acid or a nucleoprotein. To claim more, would be going beyond the experimental evidence.[1]

Taken at face value, this analysis appears quite reasonable and persuasive, but it seemed to us to ignore some of the implications of our experimental work. Our initial crude extracts contained substantial amounts of protein, but as we removed it during purification to a point where it was no longer detectable there was no loss of transforming activity. As a matter of fact, the specific activity—that is, the activity per unit weight—actually increased. Therefore, any remaining protein in the final preparations, in addition to being totally resistant to protein-splitting enzymes (even after it had been subjected to the denaturing effect of deoxycholate and the process of shaking with chloroform), had to be completely insensitive to the procedures we used for deproteinization. This all stood in sharp contrast to the exquisite sensitivity of the transforming activity to DNase. Moreover, we had the evidence that transforming activity moved with the DNA in both the ultracentrifugal and electrophoretic fields, which put further constraints on the notion that a protein present in small amounts could be responsible for the specificity.

Mirsky had an opportunity to summarize his views before a different audience at the Cold Spring Harbor Symposium that was held in June 1947 on the subject of "Nucleic Acids and Nucleoproteins." This summary came in discussion of a paper presented by Andre Boivin of Strasbourg, France, who described his recent findings on transformation of another microorganism, the common colon bacillus. The data appeared to be fully confirmatory of our pneumococcal work, although Boivin's studies later suffered some loss of credibility when other laboratories were unable to reproduce his results. It must be said for Boivin, however, that he was much less restrained in discussing the genetic implications of his findings than we had been. At his Cold Spring Harbor meeting, he made several explicit statements like the following: "In bacteria—and, in all likelihood, in higher organisms as well—each gene has as its specific constituent not a protein but a particular desoxyribonucleic acid which, at least under certain conditions (directed mutations of bacteria), is capable of functioning *alone* as the carrier of hereditary character; therefore, in the last analysis, each gene can be traced back to a macromolecule of a special desoxyribonucleic acid."[2] This was certainly a fine statement of what we believed but had been too reticent to say. Comments of this kind must have galvanized Mirsky into action, and he rose during the discussion period to give a detailed analysis, ticking off the several points of evidence for the specificity of DNA and then rebutting each of them at length. He then reiterated his conclusion that "it would be going beyond the experimental facts to assert that the specific agent in transforming bacterial types is a desoxyribonucleic acid." Boivin in reply acknowledged that one could not say this with absolute certainty but again echoed my opinion by saying that "it seems to us that the burden of the proof rests upon those who would postulate the existence of an active protein lodged in an inactive nucleic acid."

I believe that Mirsky's views must have influenced a number of people, especially the classical biologists and geneticists, but there is no way of accurately assessing the effect of this. I have been told by Swedish friends that they feel it was an important factor in the deliberations of the Nobel Prize committee in those early years after publication of our papers when Avery was still alive. The official line on this point as expressed in the Nobel Foundation's volume, *Nobel: The Man & His Prizes*, tends to support that interpretation. The discussion of our work states in part: "The discovery, because of its far-reaching implications, aroused much interest, and Avery was proposed for a Nobel Prize. But doubts were also expressed, and the Nobel Committee found it desirable to postpone an award. Actually, Avery's finding was not accepted in all quarters until A. D. Hershey . . . and M. Chase, in 1952, demonstrated that bacteriophage-DNA carries the viral genetic information from parent to progeny."[3] They do not comment on the fact that Avery lived for three years after 1952.

I do not know precisely when or on what basis Mirsky himself finally accepted that DNA was almost certain to be the carrier of genetic information. It is not likely, however, that it depended very heavily on the Hershey-Chase experiment mentioned in the Nobel book, since by its nature this was not likely to be any more convincing to him than our work was. Another line of investigation that had a bearing on the DNA thesis had been undertaken by Boivin, Vendrely, and Vendrely in France shortly after they reported their results on *Escherichia coli* transformation. They measured the amount of DNA *per* cell in various organs of the animal body and found it to be a constant value, the same for all somatic cells, but just twice the amount present in the germ cells of the same species.[4] This is just what one would expect if DNA were the genetic material, because germ cells have only half the complement of chromosomes of somatic cells. With his colleague, Hans Ris, Mirsky carried out similar studies, published in 1949, that in general supported the French results. From this they were at least willing to conclude that DNA "is part of the genic material. This does not mean, however, that the gene consists of nothing but nucleic acid."[5] This was not a major change from the Mirsky assessment of the pneumococcal transforming evidence. In the end, it must have been a combination of various pieces of evidence that persuaded him. In 1968 he wrote a piece on "The Discovery of DNA" for *Scientific American* that was chiefly a report of the early history following the original description of DNA by Miescher in 1869. He indicated in his introductory paragraph that he would "tell something of the investigative history of DNA until, some 25 years ago, it was conclusively shown to be the genetic material." That would be 1943! In the final paragraph he comes back to this point, writing that "investigators at the Rockefeller Institute—following a line of investigation with a different historical background—found that hereditary traits could be transmitted from one strain of bacteria to another by the transfer of DNA. Nucleic acid was thus shown to be the genetic material."[6] This would appear to be a substantial revision of his earlier views.

Some of Fess's old friends found it difficult to forgive Mirsky for what they conceived as his role in retarding recognition of the DNA work. I can recall an occasion when one of them would have happily prevented Mirsky from being given the honor of being selected as a Harvey Lecturer if he had been able, despite the fact that the high quality of his research clearly justified his selection. Colin MacLeod was not quite so vindictive, although he obviously was not happy about the situation. He wrote me as follows on September 16, 1958:

Dear Mac

I suppose you saw the DNA article in the Times of Sunday, September 7, which by and large was very good. I could not refrain from a snort, however, when in the second to the last paragraph Drs. Mirsky and Ris are credited with demonstrating that DNA determines the transmission of heredity. I suppose this is called poetic justice or injustice. . . . See you soon.

Colin

As far as I was concerned, I was in the position of being on the same faculty with Mirsky for the remainder of his life, and it made no sense to continue to behave as though we did not know each other. In the end, we arrived at a congenial relationship, even though one could hardly say that we were close friends.

In the course of 1945 and 1946 there were enough indications from a variety of sources of acceptance of the implications of the 1944 paper to give us some encouragement. For example, in the January 1945 issue of *American Scientist* G. Evelyn Hutchinson of Yale University included a review of our work in his "Marginalia," a section that he contributed regularly to the publication. In commenting on the significance of the findings, he noted that the transforming substance "seems to be at least a fragment of a genetic system" and concluded that "It is at any rate certain that Avery and his co-workers have made an extremely fundamental contribution not only to bacteriology and immunology, but to all the biological sciences."[7] Among the comments of bacteriologists, those of J. Howard Mueller of Harvard were especially noteworthy. After describing our findings in a 1945 review, he went on to say: "The importance of these observations can scarcely be over estimated and stimulates speculation concerning such matters as the chemical basis for specificity in nucleic acids, and the genetic implications presented by the ability to induce permanent mutation in a cell by the introduction of a chemical substance."[8]

In a similar vein, the great geneticist Sewall Wright, in a review on the "Physiological Aspects of Genetics," interpreted our work as indicating that DNA might be a chromosomal fraction acting a genetic role.[9] Sir Henry Dale went even further in his citation of Avery for the Copley Medal, which was presented by the Royal Society in 1946. He stated that pneumococcal transformation should be given "the status of genetic variation; and the substance inducing it—the gene in solution, one is tempted to call it—appears to be nucleic acid of the desoxyribose type."[10] Items of this kind certainly told us that many in biological science were interpreting our results in much the same way that we were.

At the same time that we were getting these indications that many biologists accepted our evidence, while others remained skeptical, there were also suggestions that many simply ignored it. A notable example comes from a conference on "Gene Action in Micro-organisms" that was held just one year after the publication of the 1944 paper. Not one of the several participants who presented papers on this topic mentioned pneumococcal transformation, and no discussion of the work appears in the published record of the meeting.[11] Evidence that workers in microbial genetics had not promptly incorporated the possible genetic role of DNA into their thinking was also apparent at the Cold Spring Harbor Symposium in 1946, where I gave the paper on our later work as discussed in the previous chapter. David Bonner, an important contributor to the work on the biochemical genetics of *Neurospora*, included the following comment in the summary of his paper: "There is quite general agreement at present that genes contain nucleoprotein as an essential component of their structure. One should expect, therefore, that genes, *like other proteins*, have specific configurations [italics mine]."[12] In effect, it is possible to find a wide range of responses during these early years to our claim that the pneumococcal transforming substance was DNA. The obvious conclusion is that there was no general consensus.

The most convincing evidence of acceptance of our thesis came from those scientists who based their own research on the assumption that it must be correct, even though we were of course not aware of this until somewhat later. Preeminent among these was Erwin Chargaff of Columbia University, who has acknowledged on more than one occasion that it was the 1944 paper that led him to change the course of his laboratory work and turn all of his efforts to the study of nucleic acids.[13] He began with the view that our deduction that DNA carried biological specificity meant that DNAs must differ chemically from one another, and proceeded to carry out a series of experiments showing that DNA from different species had widely varying compositions as reflected by the proportions of the four bases present in the molecule, laying to rest forever the notion that nucleic acids are all alike. As he pursued these studies further, he discovered a phenomenon that is referred to as "base complementarity": in all of the DNA samples examined, regardless of differences in overall composition, the number of molecules of the base, adenine, was equal to that of the base, thymine; and the other two bases, guanine and cytosine, were also present in equal amounts. This discovery of the base pairing of adenine-thymine and guanine-cytosine was of prime importance for understanding the structure of DNA, and it proved to be a decisive factor in the formulation of the Watson-Crick model.

There is also little reason to doubt that the work of Watson and Crick was directly influenced by the findings with pneumococcal transforming DNA, although our studies were not referred to in their papers. Watson makes this clear in *The Double Helix* in his description of the basis for Francis Crick's interest in DNA. He wrote:

> Given the fact that DNA was known to occur in the chromosomes of all cells, Avery's experiments strongly suggested that future experiments would show that all genes were composed of DNA. If true, this meant to Francis that proteins would not be the Rosetta Stone for unraveling the true secret of life. Instead, DNA would have to provide the key to enable us to find out how the genes determined, among other characteristics, the color of our hair, our eyes, most likely our comparative intelligence, and maybe even our potential to amuse others.
>
> Of course there were scientists who thought the evidence favoring DNA was inconclusive and preferred to believe that genes were protein molecules. Francis, however, did not worry about these skeptics."[14]

When they started their work on the structure of DNA in 1951, there was very little else on the record that could have turned their attention in this direction. They certainly could not have been influenced by the experiments of Hershey and Chase with bacteriophage, referred to in *Nobel: The Man & His Prizes* as providing the evidence that finally convinced all skeptics of the genetic role of DNA, since this work did not appear until 1952.

The Hershey-Chase experiments were ingenious and interesting, but there has been some debate as to whether they were responsible for the kind of impact implied in the Nobel volume. As a matter of fact, it seems likely that the reverse may have been true to some extent and that their work might not have been so readily accepted had not the information on pneumococcal DNA already existed. They obtained evidence that DNA is the genetic material using an entirely different system and different methodology, thus complementing our work very effectively. The experiments were done with bacteriophages (bacterial viruses) that were composed of roughly equal parts of protein and DNA, and each component was labeled during production of the phage with a radioactive isotope—sulfur for the protein and phosphorus for the DNA. Using these labeled phages, they obtained data which indicated that only the DNA entered the bacterial cell at the time of infection; the protein coat was left outside.[15] Thus, they provided the first clear evidence that the nucleic acid of virus carries its genetic message, a conclusion that was inescapable if one accepted the general implications of the pneumococcal work, as I had tried to point out in 1945.

Despite my change in laboratory and research activity in 1946, I did not immediately cease my attempts to serve as a missionary for the DNA story. Among the talks that I gave during the next few years, I have particularly vivid recollections of one included in a one-day "Symposium on Cancer" at the University of Illinois College of Medicine in Chicago in December 1948. This was held in honor of Professor Otto Warburg, one of the giants of German biochemistry, and was designed in part to present to him some of the recent scientific developments that he may not have been familiar with because of the information blackout during the war. I used the same title as on several previous occasions—for example, in the Lilly Award lecture—but talked principally from notes, writing out only a brief introduction and summary. The reason that I remember this occasion so well is that Warburg seemed to be barely listening to my talk and it was clear that it made no impression on him whatsoever. Again, I came away with the feeling that I had accomplished little or nothing in the way of spreading the word.

The most amusing episode in connection with these later talks occurred at my alma mater, Johns Hopkins, where I had been invited to speak at the regular evening meeting of the Johns Hopkins Medical Society in February 1949. I prepared a new manuscript, rather than warming over previous reviews of the story of pneumococcal transformation, and included some of the recent data that had been published by others. On the day of the talk, in order to have time for a visit to my old haunts at the Harriet Lane Home, I took the train to Baltimore at about noon. In the more than eight years since I had left Hopkins, there had been many changes at Harriet Lane and I found very few of my old colleagues on the scene. As a result, the visit was brief and not very conducive to relighting my nostalgic memories of the place. Through a mix-up, I had not been notified that there was a dinner for the speakers at the Faculty Club on the Homewood campus of the university, and I repaired to a restaurant across from the hospital for an early dinner.

In order to stretch out the dinner hour, I bought a copy of a Baltimore evening paper to peruse at the table. Prominently displayed on the front page, with banner headlines, was a news release from the U.S. Army announcing the discovery of a new drug for the treatment of sea sickness. This dealt with Dramamine, which had originally been introduced as an antihistamine. Dr. Leslie N. Gay, the head of the allergy clinic at the Hopkins hospital, had noted its effect on motion sickness and engaged in extensive tests of its effectiveness on sea sickness on the ships returning our troops from Europe after the war. The newspaper article ended with the information that the work would be presented that evening at 8:15 at the meeting of the Johns Hopkins Medical Society in the Hurd Memorial Hall of the Johns Hopkins Hospital.

Hurd Hall was a large amphitheater, which one entered at the rear and then moved down the steps to reach the seats that stretched on both sides of the aisle in descending tiers. There was, in addition, an ample balcony at the rear and along both sides where extra seats could be placed. By the time I reached the hall, it was thoroughly packed with people; all the seats were filled, many of the steps had been occupied, and there were numerous standees all around the balcony. In my seven years at Hopkins, I had never seen Hurd Hall so overfilled. There were only two papers on the program, the first being Dr. Gay's presentation of the impressive data on the prevention of motion sickness with Dramamine. After a short period of questions and discussion following his paper, the president of the Society got up to introduce me as the second speaker. Very little that he said could be heard because of the noise created by people streaming out of the hall. When the exodus was complete, after I had given the first few minutes of my talk, I counted approximately thirty-five hardy souls who remained in the audience because they wanted to hear about pneumococcal transformation or because they felt they had to remain out of courtesy. It is obvious that in 1949 DNA was no match for sea sickness in attracting a medical audience.

Although this episode was a source of embarrassment for some of the members of the Hopkins Medical Society, I remember being amused by it and taking the whole affair with equanimity. By that time, I no longer hoped for any immediate and dramatic signs of universal recognition of the implications of the DNA work. On the other hand, I was convinced that there was continuing progress in this direction and that the ultimate acceptance of the genetic role of DNA was only a matter of time. Perhaps this came more slowly than it might have, but I have never been attracted to the notion that our discovery was "premature." This idea was advanced by Gunther Stent in 1972 in an article in *Scientific American* entitled, "Prematurity and Uniqueness in Scientific Discovery." He introduced his thesis in the following way:

> Five years ago I published a brief retrospective essay on molecular genetics, with particular emphasis on its origins. In that historical account I mentioned neither Avery nor DNA-mediated bacterial transformation. My essay elicited a letter to the editors by a microbiologist, who complained: "It is a sad and surprising omission that . . . Stent makes no mention of the definitive proof of DNA as the basic hereditary substance by O. T. Avery, C. M. MacLeod and Maclyn McCarty. The growth of [molecular genetics] rests upon this experimental proof . . . I am old enough to remember the excitement and enthusiasm induced by the publication of the paper by Avery, MacLeod and McCarty. Avery, an effective bacteriologist, was a quiet, self-effacing, non-disputatious gentleman. These characteristics of personality should not [cause] the general scientific public . . . to let his cause go unrecognized."
>
> I was taken aback by this letter and replied that I should indeed have mentioned Avery's 1944 proof that DNA is the hereditary substance. I went on to say, however, that it is not true that the growth of molecular genetics rests on Avery's proof. For many years that proof actually had little impact on geneticists. The reason for the delay was not that Avery's work was unknown to or mistrusted by geneticists but that it was "premature."[16]

I would agree that geneticists found no immediate way to apply the information on DNA to their own experimental work, but it is hardly irrelevant that this information was directly responsible for the work of Hotchkiss (who had soon, by further purification and analytical studies on transforming DNA, convinced most biochemists, at least, that the notion of a contaminating protein was a myth), Chargaff, Watson and Crick, as well as others, which bridged the gap and made possible the development of molecular genetics. I would argue that the discovery was not "premature" but rather required further biological, chemical, and structural development before it could be manipulated by the geneticists. Since these next steps stemmed directly from the proof that DNA is the hereditary substance, the emergence of molecular genetics still has its origins in that discovery.

I cannot say that I am any more enamored of the point of view set forth by H. V. Wyatt in an article entitled, "When Does Information Become Knowledge?",[17] in which he uses our 1944 paper and its recognition as the text for his discussion. His essential point is that the information in our paper, with its low-key presentation, could not be readily fitted into accepted ideas at the time it was published and was therefore not transformed into "knowledge." Wyatt's analysis appeared before Stent's, but he returned to the topic later in a second article that considers both points of view and some of the objections that had been raised to their theses. He brings their ideas together by stating that "Thus we may extend Stent's use of 'premature' and my use of 'knowledge and information' . . . if we include a new concept: discovery can be premature if it is not capable of being extended experimentally because of technical reasons."[18] While this is almost certainly true, I don't believe that he establishes that it applies to the work on pneumococcal DNA.

At the outset of his discussion, Wyatt poses the question: "Why was it that Griffith's discovery of transformation led eventually to DNA, whereas Avery's demonstration of DNA as the genetic material did not lead immediately to a new paradigm?" He proceeds by suggesting that the answer to this assumed paradox involved technical problems: "Griffith's discovery of transformation was extended because this was technically feasible, and in the next 10 years steady progress was made by Avery's associates at the Rockefeller Institute." I would hope that I have made it clear in these pages that this statement does not very accurately reflect the state of the research during this period and that one cannot assume, as Wyatt does, that "each step brought Avery nearer to the identity of transforming principle." In contrast to this view of the research between 1928 and 1944, Wyatt felt that difficulty in following up on the 1944 paper "lay in translating the information into a form which could lead to further experiments and could be assimilated into a meaningful scientific paradigm." He sums up with the comment: "Thus Avery's discovery was premature because the technical means were not yet available to extend the work into other systems and confirm the universal nature of the phenomenon."

As I see it, the major flaw in this kind of analysis is that it assumes some sort of parallelism between the scientific approaches required to determine the chemical nature of the transforming substance on the one hand and to capitalize on the information that DNA was the carrier of genetic information on the other. One obvious next step, as we recognized early in the game, was to learn something of the chemical basis for the biological specificity of the nucleic acids. It certainly must count as an "extension" of the information on pneumococcal DNA that Chargaff began his work on the chemical composition of DNA as a result of reading our paper and that the information was also the basis for Watson and Crick's focus on the structure of the molecule. Thus, along one of the most important lines of research to be pursued, the response was prompt and effective. Wyatt also takes a different view than I do about what he calls the "extension of bacterial transformation to other systems." He notes that this did not happen until almost twenty years after the publication of Griffith's paper, but it is equally true that during this period very few labs had even been concerned with the confirmation of Griffith's results and only one with the nature of the transforming substance. On the other hand, the publication of our results on DNA soon stimulated a clear expansion of the field. Boivin's work followed shortly upon ours, and by 1950 Hattie Alexander had succeeded in obtaining DNA-mediated transformation with another bacterial species, *Hemophilus influenzae.*[19] In addition, several pieces of work soon came along demonstrating the DNA transformation of genetic characters other than capsule formation (Harriett Taylor on colony variants,[20] Robert Austrian on a type-specific protein,[21] Rollin Hotchkiss on penicillin resistance,[22] and Hattie Alexander on streptomycin resistance[23]), making it clear that the phenomenon was not restricted in some way to control of the synthesis of capsular polysaccharides. Within ten years after the publication of the 1944 paper, there was thus a great deal to show for the follow-up on the original suggestion that DNA is the carrier of genetic information. Progress only seems slow when compared to current developments in molecular genetics, but it has taken a vast amount of research in hundreds of laboratories across the world to bring us to our present capacity to manipulate genes almost at will. The earlier studies showing that the transforming principle was DNA and those that led to general recognition of the genetic role of DNA were by contrast carried out by a mere handful of investigators.

By 1970 molecular biology was flourishing and just about to be expanded fantastically by the technical developments that made possible the powerful approaches classed in the category of recombinant DNA. In the 1970s a number of books began to appear that dealt with the history of DNA and the recognition of its role as the genetic material. One of the first and perhaps most scholarly of these was written by Robert Olby, an English historian of science, who called his book *The Path to the Double Helix.* The London edition of this volume first appeared in 1973, but the edition that was available in the United States, published by the University of Washington Press, did not appear until a year later.[24] Olby had been working on the book for some years, however, and he was thus able to interview Colin MacLeod personally in the early stages of the writing and to have further correspondence with him prior to Colin's death in 1972. Olby also interviewed me at about the same time that he did Colin in 1968, and we later exchanged several letters. As a result of his contacts with the two of us, he was able to present an authoritative treatment of the pneumococcal DNA work, including a few vignettes that were not extractable from the published work. His book is an extensive history that spans the period from the first discovery of nucleic acids through the establishment of the structure of DNA by Watson and Crick.

Another of these historical volumes wirtten by Franklin H. Portugal and Jack S. Cohen, appeared in 1977. Their book, *A Century of DNA: A History of the Discovery of the Structure and Function of the Genetic Material,*[25] carries the subject through the solving of the genetic code. The authors had reviewed the annual reports to the Scientific Board of Directors in the Rockefeller University archives, and they were the first to point out that there was a hiatus in the work on transformation in the Avery laboratory between 1937 and 1940, a fact that I discovered independently in reviewing this material. They also rejected Stent's categorization of the pneumococcal DNA discovery as "premature" for much the same reason that I have above.

The most recent of these assessments of the biological revolution is Horace F. Judson's massive tome, *The Eighth Day of Creation,*[26] which is broader because it includes sections on RNA and protein in addition to the primary one on DNA. The latter section had the unusual distinction for a piece of science history of appearing first in the *New Yorker.* Judson, like Olby, had spent several years collecting material for his book, and in the course of this time interviewed most of the major scientists who figured prominently in the development of molecular biology. For reasons that are not entirely clear, he never made any effort to see MacLeod, who was still alive in the early stages of his research for the book, or to contact me. Inevitably, therefore, his treatment of the pneumococcal DNA portion of the story has a more second-hand quality and does not incorporate the view of those that were involved directly in the research. He is one of those who focuses almost entirely on the 1944 paper, and there is no reference to the subsequent work on DNase and its influence on the final proof. As a matter of fact, the book is almost devoid of reference to DNase in any connection.

It was from Olby's book that I got the first intimation that Colin MacLeod had views about my participation in the work that he had never expressed to me. Olby quotes from a letter wirtten to him by MacLeod in 1967 as follows:

By the time McCarty joined us we were virtually certain of what we were dealing with, both on the basis of the methods of preparation, the physical-chemical properties, and the elementary analysis. Moreover, we had pretty good evidence that the enzyme which destroyed activity was DNase from a variety of lines of approach. . . . Maclyn McCarty was a great help in tying things down and in getting further evidence that the enzyme was indeed DNase through purification of the enzyme from pancreas.[27]

I have to confess that I was upset when I first read this in 1975, three years afer Colin's death. It certainly doesn't jibe with my perception of what went on during those early years, and nothing that I encountered subsequently on reviewing the notes and other materials tends to support it. I can conceive of Fess and Colin keeping from me at the outset their conviction that they were dealing with DNA so as not to start me off with preconceived notions, but there would have been little reason to persist in this deception for the next several months. The data in the laboratory notes suggest that Colin's view resulted from a trick of memory in which he merged the recollection of early events with those that occurred after I had begun to work with Avery in the fall of 1941.

First, there was no preparation of purified pneumococcal DNA, separated from the bulk of RNA and SSSIII, that would have been useful for elementary analysis and physical-chemical studies until the fall of 1942. One elementary analysis is to be found recorded in the notes before this time, on May 26, 1941, and it serves to make my point that the material then available was not suitable for this kind of analysis. The sample contained more carbon but less than half the amount of phosphorus and nitrogen found in our later preparations or in authentic DNA, suggesting that a large component of SSSIII was present. With respect to the evidence that the enzymes that destroyed transforming activity were DNase, I have been able to find no record of these enzymes being tested on DNA until I did it in the summer of 1942. In fact, no method for measuring DNase activity had been available in the lab.

I feel, therefore, that I must stick with the interpretation of the events as I have unfolded them in these pages. My conclusion that MacLeod's memory was playing tricks on him receives some support from another Olby quote, this time from his recorded interview with Colin. In reporting on a visit that he and Avery made to P. A. Levene to discuss transformation, Colin says: "He [Levene] was skeptical about the possible role of DNA in transformation reactions." The difficulty here is that Levene died in September 1940, and it was not until late January 1941 that Avery and MacLeod had discovered, rather to their surprise, that their extracts contained a small amount of DNA. They could not have raised the issue of DNA with Levene on the basis of any specific information about the pneumococcal material.

It has been my purpose to tell the story of the discovery of the genetic role of DNA, along with what I feel to be the necessary background, and I will not dwell further on the aftermath. Certainly this is no place to attempt an assessment of the current status of the explosive developments in DNA science. Because my research has followed a different path since those early years, I have not been personally involved in any of these developments since the late 1940s, even though I have followed them with interest. It has been an exciting time for biologists in all branches of the field, including medicine.

The stimulus for finally getting down to the business of writing this story came from the celebration of the thirty-fifth anniversary of the publication of the 1944 paper. The talk that I prepared on this occasion sent me back to the old laboratory notes and archives, and I realized that a more serious job of researching the available material would have to be carried out in order to piece together the different phases of the investigation. Joshua Lederberg originated the idea for this anniversary celebration. After he had accepted the appointment as the new President of the Rockefeller University, but before arriving on the scene from Stanford, he sent me a handwritten memo dated February 1, 1978:

A 35th Anniversary coming

Dear Mac—

I was just noticing the date and reflecting that it was the 34′ birthday of *the* DNA paper.

2/1/44 was an important day in my life, and it would disappoint me to let 2/1/79 go by without notice. I really would look forward to having a convocation planned around that theme—perhaps a supplement to J. Exp. Med., or whatever, as well. . . . And it's time you brought your own reminiscences out of the closet! I'll ask you what you've thought about it in March (visit). . . .

Yours, Josh

I had known Lederberg since the Cold Spring Harbor Symposium in 1946, which he had attended as a very active and articulate graduate student. He gives the 1944 paper credit for changing the course of his career, moving him out of medical school into the graduate study of genetics; and he was one of those who had quite soon applied the information in the paper to his own laboratory work in an unsuccessful attempt with F. J. Ryan to cause mutations in the mold *Neurospora* with DNA. I was quite sympathetic with his idea of celebrating the thirty-fifth anniversary but did not think it was appropriate for me to take a leading role in arranging it. The matter was not forgotten, however, and plans were put in operation that fall for a special colloquium at the University on February 2, 1979. Lederberg, Dubos, Hotchkiss, and I all spoke, and each member of the audience was provided with a reprint of the original paper which had been reproduced, together with a foreword by Josh, in the February issue of the *Journal of Experimental Medicine*.

I gave my talk at the anniversary meeting the title: "The

Identification of the Pneumococcal Transforming Substance as DNA: A Retrospective Look at How We Got There." It was not published, but I used it extensively in preparing an invited historical paper for the *Annual Review of Genetics* on "Reminiscences of the Early Days of Transformation."[28] These efforts moved me to think in terms of doing a more thorough and complete job, which now, for better or worse, is finished.

BIBLIOGRAPHY

THE MANY REFERENCES throughout the text to the Annual Scientific Reports to the Board of Scientific Directors of the Rockefeller Institute for Medical Research are indicated by the date on which the reports were submitted and they are not cited in this section. In addition, the material drawn from the laboratory notes is also identified by date. The annual reports are to be found in the Rockefeller University Archives, and it is my intention to deposit the laboratory notes in this same resource as soon as I have finished working with them. The present section is designed primarily to provide the specific reference citations for the published work to which I have referred.

A comprehensive treatment of all of the research on pneumococci through the mid-1930s was provided in a volume prepared by Avery's old friend and colleague at the Hoagland Laboratory, Benjamin White (*The biology of pneumococcus*. 1938. New York: The Commonwealth Fund). This book was recently reprinted (1979) by Harvard University Press so that it remains available to modern students of the subject.

I. THE PREPARATORY YEARS

1. Luck, J.M. 1936. Liver proteins. I. The question of protein storage. *Journal of Biological Chemistry* 115:491–510.

II. THE MEDICAL SCENE

1. Scott, D.A., and A.F. Charles. 1933. Studies on heparin. III. The purification of heparin. *Journal of Biological Chemistry* 102:437–448.

2. Hodes, H.L., W.C. Stifler, Jr., E. Walker, M. McCarty, and R.G. Shirley. 1939. The use of sulfapyridine in primary pneumococcic pneumonia and in pneumococcic pneumonia associated with measles. *Pediatrics* 14:417–446.

3. McCarty, M. 1941. Effect of *p*-aminobenzoic acid on therapeutic and toxic action of sulfapyridine. *Proceedings of the Society for Experimental Biology and Medicine* 46:133–136.

4. McCarty, M., and W.S. Tillett. 1941. The inactivating effect of sulfapyridine on the leukotoxic action of benzene. *Journal of Experimental Medicine* 74:531–544.

III. THE SUGARCOATED MICROBE

1. Neufeld, F., and L. Händel. 1910. Weitere Untersuchungen über Pneumokokken-Heilsera. III. Mitteilung. Uber Vorkommen und Bedeutung atypischer Varietäten des Pneumokokkus. *Arbeiten aus dem kaiserlichen Gesundheitsamte* 34:293–304.

2. Dochez, A.R., and L.J. Gillespie. 1913. A biologic classification of pneumococci by means of immunity reactions. *Journal of the American Medical Association* 61:727.

3. This story, which I heard from Avery on more than one occasion, is also told by Dubos in his scientific biography of Avery (Dubos, R.J. 1976. *The Professor, the Institute, and DNA*, 63. New York: The Rockefeller University Press).

4. Dochez, A.R., and O.T. Avery. 1917. The elaboration of a specific soluble substance by pneumococcus during growth. *Journal of Experimental Medicine* 26:477–493.

5. Heidelberger, M. 1977. A "pure" organic chemist's downward path. *Annual Review of Microbiology* 31:1–12 (quote on p.9).

6. Ibid., 10.

7. Heidelberger, M., and W.F. Goebel. 1926. The soluble specific substance of pneumococcus. IV. On the nature of the specific polysaccharide of Type III pneumococcus. *Journal of Biological Chemistry* 70:613–624.

8. Goebel, W.F. 1939. Studies on antibacterial immunity induced by artificial antigens. I. Immunity to experimental pneumococcal infection with an antigen containing cellobiuronic acid. *Journal of Experimental Medicine* 69:353–364.

9. Avery, O.T., and R. Dubos. 1931. The protective action of a specific enzyme against type III pneumococcus infection in mice. *Journal of Experimental Medicine* 54:73–89.

IV. TRANSFORMATION

1. Griffith, F. 1928. The significance of pneumococcal types. *Journal of Hygiene* 27:113–159 (quote on p. 117).
2. Ibid., 129–130.
3. Ibid.
4. Ibid., 153.
5. Neufeld, F., and W. Levinthal. 1928. Beiträge zur Variabilität der Pneumokokken. *Zeitschrift für Immunitätsforschung* 55:324–340.
6. Dawson, M.H. 1928. The interconvertibility of "R" and "S" forms of pneumococcus. *Journal of Experimental Medicine* 47:577–591.
7. Dawson, M.H. 1930. The transformation of pneumococcal types. II. The interconvertibility of type-specific pneumococci. *Journal of Experimental Medicine* 51:123–147.
8. Dawson, M.H., and R.H.P. Sia. 1931. In vitro transformations of pneumococcal types. I. A technique for inducing transformation of pneumococcal types in vitro. *Journal of Experimental Medicine* 54:681–700.
9. Sia, R.H.P., and M.H. Dawson. 1931. In vitro transformations of pneumococcal types. II. The nature of the factor responsible for the transformation of pneumococcal types. *Journal of Experimental Medicine* 54:701–710.
10. I do not know what publication Francis is referring to here.
11. This talk was published in *International Virology I. Proceedings of the First International Congress of Virology,* 1969, pp. 224–228. Basel: S. Karger.
12. Alloway, J.L. 1932. The transformation in vitro of R pneumococci into S forms of different specific types by the use of filtered pneumococcus extracts. *Journal of Experimental Medicine* 55:91–99.
13. Alloway, J.L. 1933. Further observations on the use of pneumococcus extracts in effecting transformation of type in vitro. *Journal of Experimental Medicine* 57:265–278.
14. Letter to O.T. Avery from Simon Flexner 6 October 1931. From the Papers of Simon Flexner in the archives of the American Philosophical Society.

V. ENTER MACLEOD

1. Sevag, M.G. 1934. Eine neue physikalische Enteiweissungsmethode zur Darstellung biologisch wirksamer Substanzen. Isolierung von Kohlenhydraten aus Hühnereiweiss und Pneumokokken. *Biochemische Zeitschrift* 273:419.

VIII. INTIMATION OF SUCCESS

1. MacLeod, C.M., and M. McCarty. 1942. The relation of a somatic factor to virulence of pneumococci [Abstract]. *Journal of Clinical Investigation* 21:647.

2. Caspersson, T., E. Hammarsten, and H. Hammarsten. 1935. Interactions of proteins and nucleic acids. *Transactions of the Faraday Society* 31:367–389.

3. Schultz, J. 1941. The evidence of the nucleoprotein nature of the gene. *Cold Spring Harbor Symposia on Quantitative Biology* 9:56–65.

4. Gortner, R.A. 1929. *Outlines of Biochemistry*, 358, 399. New York: John Wiley & Sons.

5. Leathes, J.B. 1926. Function and design. *Science* 64:387–394.

6. Mirsky, A.E., and A.W. Pollister. 1946. Chromosin, a desoxyribose nucleoprotein complex of the cell nucleus. *Journal of General Physiology* 30:121.

IX. THE HOME STRETCH

1. The original of this historic letter is now among the O.T. Avery Papers in the Tennessee State Library and Archives, Nashville, Tennessee. The complete text is available, including the first three pages which were completed before Avery decided to undertake the description of his current work in the laboratory.

2. Bresch, C. 1964. *Klassische und molekulare Genetik*, 130. Berlin: Springer-Verlag.

3. Hotchkiss, R.D. 1966. "Gene, transforming principle, and DNA." In *Phage and the Origins of Molecular Biology*, edited by J. Cairns, G.S. Stent, and J.D. Watson, pp. 185–187. New York: Cold Spring Harbor Laboratory of Quantitative Biology.

4. Dubos, R.J. 1976. *The Professor, the Institute, and DNA*, Appendix I, pp. 217–220. New York: The Rockefeller Press.

5. Letter from H.A. Schneider to President Detlev W. Bronk of Rockefeller University written on September 22, 1965, with reference to the dedication of the Avery Memorial Gateway, which took place at the University on September 29. The letter is from the Detlev Bronk Papers in the Rockefeller University Archives.

6. Avery, O.T., C.M. MacLeod, and M. McCarty. 1944. Studies on the chemical nature of the substance inducing transformation of pneumococcal types. Induction of transformation by a desoxyribonucleic acid fraction isolated from pneumococcus type III. *Journal of Experimental Medicine* 79:137–158.

X. STRENGTHENING THE EVIDENCE

1. Reported in G.W. Corner. 1964. *A History of the Rockefeller Institute, 1901–1953: Origins and Growth*, 430. New York: The Rockefeller Institute Press.

2. McCarty, M. 1980. Reminiscences of the early days of transformation. *Annual Review of Genetics* 14:1–15.

3. McCarty, M. 1945. Reversible inactivation of the substance inducing

transformation of pneumococcal types. *Journal of Experimental Medicine* 81:501–514.

4. Fischer, F.G., I. Böttger, and H. Lehmann-Echternacht. 1941. Über die Thymo-polynucleotidase aus Pankreas. Nucleinsäure. V. *Zeitschrift für physiologisches Chemie* 271:246–264.

5. McCarty, M. 1946. Purification and properties of desoxyribonuclease isolated from beef pancreas. *Journal of General Physiology* 29:123–139.

6. McCarty, M., and O.T. Avery. 1946. Studies on the chemical nature of the substance inducing transformation of pneumococcal types. II. Effect of desoxyribonuclease on the biological activity of the transforming substance. *Journal of Experimental Medicine* 83:89–96.

7. McCarty, M., and O.T. Avery. 1946. Studies on the chemical nature of the substance inducing transformation of pneumococcal types. III. An improved method for the isolation of the transforming substance and its application to pneumococcus types II, III, and VI. *Journal of Experimental Medicine* 83:97–104.

8. Kunitz, M. 1948. Isolation of crystalline desoxyribonuclease. *Science* 108:19–20.

9. Kunitz, M. 1950. Crystalline desoxyribonuclease. I. Isolation and general properties. Spectrophotometric method for the measurement of desoxyribonuclease activity. II. Digestion of thymus nucleic acid (desoxyribonucleic acid). The kinetics of the reaction. *Journal of General Physiology* 33:349–377.

10. McCarty, M. 1947. The occurrence during acute infections of a protein not normally present in the blood. IV. Crystallization of the C-reactive protein. *Journal of Experimental Medicine* 85:491–498.

XI. THE FINAL MONTHS

1. Wyatt, H.V. 1975. Knowledge and prematurity: The journey from transformation to DNA. *Perspectives in Biology and Medicine* 18:149–156.

2. *Science* October 27, 1944, 100:375.

3. *Science* October 13, 1944, 100:328–329.

4. Stanley, W.M. 1970. The "undiscovered" discovery. *Archives of Environmental Health* 21:256–262.

5. McCarty, M. 1946. Chemical nature and biological specificity of the substance inducing transformation of pneumococcal types. *Bacteriological Reviews* 10:63–71.

6. McCarty, M., H.E. Taylor, and O.T. Avery. 1946. Biochemical studies of environmental factors essential in transformation of pneumococcal types. *Cold Spring Harbor Symposia on Quantitative Biology* 11:177–183.

7. Hotchkiss, R.D. 1966. "Gene, transforming principle, and DNA." In *Phage and the Origins of Molecular Biology*, edited by J. Cairns, G.S. Stent, and J.D. Watson, p. 188. New York: Cold Spring Harbor Laboratory of Quantitative Biology.

XII. AFTERMATH

1. Mirsky, A.E., and A.W. Pollister. 1946. Chromosin, a desoxyribose nucleoprotein complex of the cell nucleus. *Journal of General Physiology* 30 (quote on pp. 134–135).

2. Boivin, A. 1947. Directed mutation in colon bacilli, by an inducing principle of desoxyribonucleic nature: Its meaning for the general biochemistry of heredity. *Cold Spring Harbor Symposia on Quantitative Biology* 12:7–17.

3. *Nobel: The Man & His Prizes*, 3rd edition, 1972 (quote on p. 201). Edited by the Nobel Foundation and W. Odelberg. New York: American Elsevier.

4. Boivin, A., R. Vendrely, and C. Vendrely. 1948. L'acide desoxyribonucleique due noyau cellulaire, depositaire des caractères hereditaires; arguments d'ordre analytique. *Comptes rendus hebdomadaires des Séances de l'Académie des Sciences, Paris* 226:1061–1063.

5. Mirsky, A.E., and H. Ris. 1949. Variable and constant components of chromosomes. *Nature (London)* 163:666–667.

6. Mirsky, A.E. 1968. The discovery of DNA. *Scientific American* 218:78–88.

7. Hutchinson, G.E. 1945. The biochemical genetics of *Pneumococcus*. *American Scientist* 33:56–57.

8. Mueller, J.H. 1945. The chemistry and metabolism of bacteria. *Annual Review of Biochemistry* 14:734.

9. Wright, S. 1945. Physiological aspects of genetics. *Annual Review of Physiology* 7:75–106.

10. Dale, Sir H. 1946. Address of the president. *Proceedings of the Royal Society (London)* 185A:128.

11. *Annals of the Missouri Botanical Garden*, 1945. Special Number, Conference on "Gene Action in Micro-organisms," 32:107–263.

12. Bonner, D. 1946. Biochemical mutations in *Neurospora*. *Cold Spring Harbor Symposia on Quantitative Biology* 11:21.

13. Chargaff, E. 1963. *Essays on Nucleic Acids*. New York: Elsevier. Chargaff, E. 1978. *Heraclitean Fire*, pp. 82–84. New York: The Rockefeller University Press.

14. Watson, J.D. 1968. *The Double Helix* (quote on p. 14). New York: Atheneum.

15. Hershey, A., and M. Chase. 1952. Independent functions of viral proteins and nucleic acid in growth of bacteriophage. *Journal of General Physiology* 36:39–56.

16. Stent, G.S. 1972. Prematurity and uniqueness in scientific discovery. *Scientific American* 227:84.

17. Wyatt, H.V. 1972. When does information become knowledge? *Nature* 235:86–89.

18. Wyatt, H.V. 1975. Knowledge and prematurity: The journey from transformation to DNA. *Perspectives in Biology and Medicine* 18:149–156.

19. Alexander, H.E., and G. Leidy. 1951. Determination of inherited traits of *H. influenzae* by desoxyribonucleic acid fractions isolated from type-specific cells. *Journal of Experimental Medicine* 93:345–359.

20. Taylor, H.E. 1949. Additive effects of certain transforming agents from some variants of pneumococcus. *Journal of Experimental Medicine* 89:399–424.

21. Austrian, R., and C. M. MacLeod. 1949. Acquisition of M protein by pneumococci through transformation reactions. *Journal of Experimental Medicine* 89:451–460.

22. Hotchkiss, R.D. 1951. Transfer of penicillin resistance in pneumococci by the desoxyribonucleate fractions from resistant cultures. *Cold Spring Harbor Symposia on Quantitative Biology* 16:457–461.

23. Alexander, H., and G. Leidy. 1953. Induction of streptomycin resistance in sensitive *Hemophilus influenzae* by extracts containing desoxyribonucleic acid from resistant *Hemophilus influenzae. Journal of Experimental Medicine* 97:17–31.

24. Olby, R. 1975. *The Path to the Double Helix*. Seattle: University of Washington Press.

25. Portugal, F.H., and J. S. Cohen. 1977. *A Century of DNA*. Cambridge, Mass.: MIT Press.

26. Judson, H.F. 1979. *The Eighth Day of Creation*. New York: Simon and Schuster.

27. Olby, R. *The Path to the Double Helix*, p. 185.

28. McCarty, M. 1980. Reminiscences of the early days of transformation. *Annual Review of Genetics* 14:1–25.

INDEX